Innovation in Small Construction Firms

Also available from Taylor & Francis

Construction the Future: nD Systems
G. Aouad et al.   Hb: ISBN 9780415391719

Innovation in Construction: an International Review of Public Policies
A Manseau et al.   Hb: ISBN 9780415254786

Understanding Quality in Construction
H. Chung   Hb: ISBN 9780419249504

Understanding I.T. in Construction
M. Sun et al.   Pb: ISBN 9780415231909

Introduction to Building Procurement Systems 2nd ed.
J. Masterman   Hb: ISBN 9780415246415
                Pb: ISBN 9780415246422

Information and ordering details
For price, availability and ordering visit our website **www.sponpress.com**
Alternatively our books are available from all good bookshops.

Innovation in Small Construction Firms

Peter Barrett, Martin Sexton and Angela Lee

Taylor & Francis
Taylor & Francis Group

LONDON AND NEW YORK

First published 2008
by Taylor & Francis
2 Park Square, Milton Park, Abingdon, Oxfordshire OX14 4RN

Simultaneously published in the USA and Canada
by Taylor & Francis
711 Third Avenue, New York, NY 10017

First issued in paperback 2016

Taylor & Francis is an imprint of the Taylor & Francis Group, an informa business

© 2008 Peter Barrett, Martin Sexton and Angela Lee

Typeset in Times New Roman by
Keystroke, 28 High Street, Tettenhall, Wolverhampton

British Library Cataloguing in Publication Data
A catalogue record for this book is available from the British Library

Library of Congress Cataloging-in-Publication Data
Barrett, Peter, professor.
  Innovation in small construction firms/Peter Barrett, Martin Sexton,
  and Angela Lee.
      p. cm.
    Includes bibliographical references and index.
  1. Construction industry—Research.   2. Construction industry—
  Technological innovations.   3. Small business—Technological
  innovations.   I. Sexton, Martin, 1966–   II. Lee, Angela, Ph. D. III. Title.
  TH213.5.B37 2008
  690.068′4—dc22   2007024814

ISBN 13: 978-1-138-99250-4 (pbk)
ISBN 13: 978-0-415-39390-4 (hbk)

Contents

Illustrations

Figures

Tables

Acknowledgements

Figures 2.2.–2.5 are reproduced from Sexton, M. and Barrett. P., (2003), 'A literature synthesis of innovation in small construction firms: Insights, ambiguities and questions', *Construction Management and Economics*, 21: 613–622.

Figures 4.1, 5.1, 6.1, 7.1, 7.2, 9.2, 9.3, 9.4 and 9.5 are reproduced from Sexton, M. and Barrett. P., (2003), 'Appropriate innovation in small construction firms', *Construction Management and Economics*, 21:623–633.

Published by Taylor & Francis (www.informaworld.com).

Chapter I

Introduction

Characteristics of the construction industry

The construction industry is not homogeneous. It is composed of many diverse competing (and collaborating) firms, the majority of whom are brought together for one, bespoke project, before transferring to other projects. The sector is often characterised by its adversarial behaviour, litigious orientation, poor communication and coordination, lack of customer focus, and its low investment in research and development (Simon, 1944; Emmerson, 1962; Banwell, 1964; Latham, 1994; Egan, 1998; Fairclough, 2002). Within this unfavourable supply context, clients are increasing their demands for improved building performance (both functionally and aesthetically), while at the same time reducing initial capital, and ongoing operational and maintenance costs. Set against an already competitive industry, construction firms are under pressure to develop and/or adopt innovative technologies and practices in order to try to satisfy these demands (Sexton et al., 2005).

The construction industry as a whole has a poor reputation for innovation and is often accused of being slow to adopt new technologies. Further, due to the weak appropriability conditions found in construction, contractors have little to gain from being innovative other than optimising their own processes (Sexton and Barrett, 2005). Economies of scale often do not exist and knowledge gains are rarely transferred (Pries and Janszen, 1995). However, Ball (1988) has argued that the industry is not backward, merely different from other industries – it is an industry that has to 'innovate' on a daily basis in order to solve the problems that the design and production phases pose. This stance is supported by Tatum (1984, 1986), Pries and Janszen (1995) and Veshosky (1998), who affirm that construction projects are, by their very nature, inherently innovative: the project-based nature of the industry makes every project unique.

Nam and Tatum (1997) move the debate along by suggesting that the lack of innovation may not be attributed to the lack of capability, but to the absence of a coordinated effort to link market needs and inventive capacity in spite of adequate demand pull as well as a supply of promising technologies, such as computers, robotics and advance materials that are ready to be utilised through a coordinated system. Further, Reichstein et al. (2005) report that construction firms are less open to the external environment and tend to have poorly developed research and development (R&D), with low capacity to absorb practices from other sectors.

Notwithstanding if the construction sector is less innovative than others, the desire for innovation in construction is well recognised (e.g. Atkin, 1999; Manseau and Seaden, 2001). Gann (2000: 220) comments that construction firms need to improve their capabilities in managing innovation if they are to 'build reputations for technical excellence that set them apart from more traditional players'. Moreover, Sexton and Barrett (2003a: 613) remark that successful innovation enables construction firms to better satisfy 'the aspirations and needs of society and clients, whilst improving their competitiveness in dynamic and abrasive markets'. Although the demand for innovation is unquestioned, research into innovation in the context of construction is sparse. Research into innovation in construction is not specific to the industry and still very much in its embryonic stage. Winch (1998: 272) echoes this view and suggests that more innovation research is required to 'get a grip on the sources and applications of new ideas in the construction industry'.

In an effort to bring in a new 'can innovate, should innovate, want to innovate' construction industry culture, this book aims to address this research gap, aiming specifically to promote the benefits of innovation and stimulate innovation capability within and between small and medium sized construction firms who make up the majority of the industry. The results are drawn from case studies of innovation activity in seven small construction firms. The issues addressed include:

- the focus and outcome of innovation
- the organisational capabilities of innovation
- the context of innovation
- the process of innovation.

This chapter begins by setting the backdrop of the UK construction industry, around which this investigation is based, then leads on to define what a small and medium enterprise (SME) is and why innovation is imperative.

The UK construction industry

The construction industry is Europe's largest industrial employer, representing 7.2 per cent of the continent's total employment and 9.9 per cent gross domestic product (GDP) (FIEC (European Construction Industry Federation), 2003; see Table 1.1). In the UK, the industry employs 8 per cent of the workforce (Department of Trade and Industry (DTI), 2003a).

The construction industry is one of the most complex and dynamic industrial sectors. It relies heavily on skilled manual labour that is supported by an interconnected management and design input, which is often highly 'fragmented' right up to the point of delivery (Mohsini and Davidson, 1992). A large and complex project will involve many design, construction and supplier organisations, whose

Table 1.1 Number of construction employees and production by country

Country	Number employed	Production, billion euros
Europe		
Austria	258,000	28
Belgium	230,000	28
Germany	2,418,000	213
Denmark	162,000	20
Spain	1,825,000	103
Greece	320,000	14
France	1,512,000	118
Finland	146,000	17
Great Britain	1,629,000	139
Italy	1,748,000	107
Ireland	186,000	20
Luxembourg	30,000	1
Netherlands	483,000	49
Portugal	618,000	24
Sweden	215,000	24
Total	11,780,000	905
Other		
Japan	6,230,000	626
United States	6,544,000	880

Source: FEIC, 2003

sporadic involvement will change throughout the course of the project (for example, see Carty, 1995). The organisations will be both large and small, and although they have usually never met before, they are expected to work together effectively and efficiently throughout the duration of the project. Complicating this situation yet further, the significant majority of design and construction activities are subcontracted, which renders collaborative and integrated working extremely problematic. In addition, design and construction practitioners typically find themselves working on several projects at the same time. According to Mullins' (1999) generic and rather simplistic prescription, the success of a project relies heavily on having clearly defined objectives and well-defined tasks. But this is not always apparent in construction, as the client's objectives often crystallise only over the course of the project.

Table 1.2 illustrates the variety of disciplines and trades in the UK construction sector. The number of main trade organisations has fallen steadily from 84,885 organisations in 1992, to 54,043 in 2002 (DTI, 2003a). This equates to a 36.3 per cent decrease in main trade organisations compared to a 19.2 per cent decrease of the total number of organisations for the same period. Thus, the variety of specialised organisations prevails over that of their main trade counterparts. According to Pearce (2003), the variety and diversity of organisations in the construction sector is imperative due to the complex nature of the design and

Table 1.2 Number of UK construction organisations by trade per year

Trade of organisation	1992	1994	1996	1998	2000	2002
General builders	74,393	69,160	—	—	—	—
Building and civil engineering contractors	6,180	6,845	—	—	—	26,201
Non-residential building	—	—	—	—	—	13,462
House building	—	—	—	—	—	14,380
Civil engineering	4,312	4,812	—	—	—	—
Total: main trades	*84,885*	*80,187*	*66,380*	*63,550*	*59,708*	*54,043*
Construction engineers	2,713	2,168	1,216	864	1,105	—
Demolition	717	685	740	793	855	1,137
Reinforced concrete specialists	859	637	415	321	263	182
Test drilling and boring	—	—	—	—	—	—
Roofing	7,524	6,470	5,457	5,599	6,310	6,252
Asphalt and tar sprayers	1,163	1,077	866	711	845	—
Construction of highways	—	—	—	—	—	1,640
Construction of water projects	—	—	—	—	—	341
Scaffolding	1,779	1,733	1,270	1,009	1,555	1,194
Installation of electrical wiring and fitting	21,780	21,004	19,463	19,385	18,426	20,424
Insulating activities	1,265	1,131	977	934	879	993
Plumbing	14,647	13,181	11,698	12,519	13,937	18,853
Heating and ventilating engineers	9,774	9,136	6,697	5,500	5,870	—
Plastering	3,893	3,160	2,475	2,538	2,389	2,777
Joinery installation	14,199	12,614	10,202	10,016	9,699	15,295
Flooring contractors	2,387	2,320	2,249	2,684	2,820	—
Floor and wall tiling specialists	1,607	1,430	1,011	770	791	—
Floor and wall covering	—	—	—	—	—	4,058
Suspended ceiling specialists	1,757	1,509	2,118	4,529	3,452	—
Painting	10,788	8,974	8,284	8,969	8,507	7,895
Glazing	7,001	6,918	4,174	3,128	3,581	3,574
Plant hire (with operators)	5,261	5,940	4,607	3,882	3,245	6,942
Other construction work	11,345	14,383	13,016	15,535	19,154	20,578
Total: all trades	*205,704*	*194,657*	*163,315*	*163,236*	*163,426*	*166,181*

Source: Adapted from DTI, 2003a

construction of modern buildings and facilities. Conversely, in comparison, the number of persons employed by the industry has risen by 9.9 per cent due to the rise in demand for construction work; in 1992 the industry employed 1,846,000 persons and in 2002 this figure rose to 2,029,000 (DTI, 2003a). Therefore, the inverse correlation between the number of organisations (particularly main trade) against the number of persons employed would suggest that smaller specialist organisations are foremost.

Thus, the majority of the construction labour market is self-employed (Briscoe et al., 2000). The scale of small organisation activity in the UK construction industry is considerable, with, in 2002, 99.3 per cent of UK construction organisations having between one and seventy-nine staff and employing 65.4 per cent of the total construction workforce (DTI, 2003a; see Figure 1.1). According to Storey (1998) and Smith and Whittaker (1998), there is no single, uniformly acceptable, definition of a small organisation that can be applied across all industrial sectors. Definitions that relate to 'objective' measures of size, such as the number of employees, sales turnover, profitability and net worth, are not comparable across all industries. In 1971, the Bolton Committee employed a discordant portfolio of different definitions of a small organisation according to the sector in which it operated (Bolton, 1971). Bolton defined a small organisation in the construction sector as having twenty-five or fewer employees, but this made international comparisons virtually impossible. Bolton's statistical definition has not been agreed nor widely employed by construction researchers and economists. To overcome the divergence of categorisations, the European Commission (EC) coined the term 'small and medium enterprise'. The SME sector is therefore taken to be enterprises – except agriculture, hunting, forestry and fishing which employ fewer than 249 workers – and is further disaggregated into three components (European Commission, 2003):

- micro enterprises: those with 0–9 employees
- small enterprises: those with 10–49 employees
- medium enterprises: those with 50–249 employees.

Although the EC's definition of SMEs again has not been unequivocally implemented – in government-sponsored studies of UK Training and Enterprise Councils (TECs), five out of six reports defined SME as under 199 employees, and the sixth as under 250 employees (Whittaker et al., 1997) – it will be adopted throughout the course of this book.

The predominance of micro enterprises in the UK construction industry, employing between one and nine persons, over larger-sized organisations (see Figure 1.1) may be attributed to the fact that large contracts require specialist work and the specialist contractors are predominantly self-employed and, where necessary, employ a few additional hands (Abdel-Razek and McCaffer, 1987; Gale and Fellows, 1990). According to Langford and Male (1992), larger organisations generally resort to a greater use of subcontractors (micro and small enterprises) in

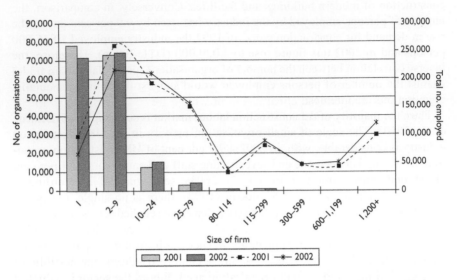

Figure 1.1   Number of UK construction organisations and employment figures
in 2001 and 2002

Source: adapted from DTI, 2003a

a bid to reduce the overhead burden of tax, National Insurance contributions and working capital needs. This stance may contribute to the greater decline of the number of main trade organisations, as evidenced in Table 1.2. Thus, any overall performance improvement of the industry through innovation will be significantly influenced by SMEs, of various trades, as they make up the majority of the industry.

The predominance of SMEs in construction is not unique as is often thought. The agriculture, hunting, forestry and fishing sectors follow a similar pattern to that of construction (SME Statistics, 2002; see Figure 1.2). The manufacturing sector is seemingly a mirror to that of construction: more persons are employed in large enterprises than micro and small enterprises. Only the 'wholesale and retail trade/repairs' and 'real estate, renting and business activities' sectors are relatively balanced in terms of the number of employed in micro and large enterprises. However, only the construction industry, with the exception of manufacturing in some instances, operates within a project-based environment with temporary teams consisting of actors from various discipline backgrounds. The structure of the industry is arguably a function of the work it is called upon to do.

Why research innovation in SMEs?

Despite increasing interest in the small business in the UK, its importance in the economy is often still underestimated (Storey, 1998). According to Curran and Blackburn (2001), the primary reason for this is that there still remains a tendency

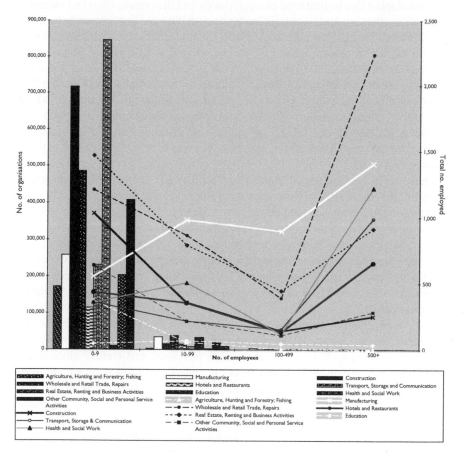

Figure 1.2   Number of UK enterprises and employment figures by industrial
              sector in 2001

Source: SME Statistics, 2002

to see small businesses as less central to economic activities than large businesses. Further, much of the research conducted so far has not been high quality due mainly to failures to recognise the special problems studying the small business poses for researchers. This book aims to address this research gap by viewing innovation in SMEs within the context of its wider role in the construction industry.

The aspiration to enhance construction performance has been traditionally checked by the industry's assumption that the intrinsic characteristics of construction and the construction industry – such as industry sector fragmentation, 'boom-and-bust' market cycles, use of relatively low technology and antagonistic procurement policies – inhibits innovation (for example, see Ball, 1988; Powell, 1995; Gann, 2000). Brouseau and Rallet (1995) capture much of this debate in their contention that institutional characteristics and the organisation and management of construction itself constrain innovation activity and restrict parties to apply innovations. Indeed, it has been argued that 'the construction industry is infamous for the barriers it places in the way of innovation' (CERF (Civil Engineering Research Foundation), 2000), and that 'the desire for [construction] firms to change has come more from a fear of being left behind by competitors than from a belief in the benefits of innovation' (Yisa et al., 1996: 49). Although it is acknowledged that construction firms have always demonstrated an ability to innovate (for example, see Slaughter, 1998), construction practitioners are now very much getting to grips with the need for, and management of, innovation as an explicit endeavour.

These objectives are linked to the assertion found in the general construction literature that innovation performance must improve, and more specifically from the research which investigated innovation in large construction firms. In the UK, public policy instruments to develop innovation in the construction industry have been categorised into programmes that support research and development; advanced practices and experimentation; performance and quality improvement; and taking up systems and procedures (CIB Task Group 35, 2000). Moreover, the relevance and accessibility of many of these initiatives for SMEs is debatable (Sexton et al., 1999).

Innovation theory and practice are being drawn from established bodies of innovation knowledge predominately based on other industries (for example, see Sexton and Barrett, 2003a), but they have not been sufficiently envisioned, embedded and evaluated in a construction context to form a robust body of construction innovation knowledge in its own right. Similarly, it is argued that '[construction] project-based, service-enhanced forms of enterprise are inadequately addressed in the innovation literature' (Gann and Salter, 2000: 955). These observations are extended further by commenting that to our knowledge the construction innovation literature often emphasises construction firms of large size, and that innovation in small firms has been generally ignored.

We neglect small construction firms at our peril, as considerable evidence from the general innovation literature indicates that there is a significant difference in the innovation capability and output of small firms compared to large firms, with it being argued, for example, that small firms are organic in nature making them more agile and responsive, while large firms tend to be more mechanistic (for example, see Mansfield et al., 1971; Rothwell, 1989; Nooteboom, 1994; Rothwell and Dodgson, 1994). This difference needs to be understood, in order to underpin policy and corporate guidance. Drawing upon similar concerns in the design of technology transfer mechanisms for construction small to medium enterprises, it has been stressed that there is a

> need to appreciate that construction SMEs and large construction companies are different animals, that live in different business market habitats, that must behave in different ways in order to adapt and succeed, and which need

different sources and types of knowledge and technology to remain nourished and healthy.

(Sexton et al., 2006: 21)

Summary

This chapter has set the scene for the need for SME innovation research in the construction industry. The context of the industry was illustrated, and the predominance of SMEs in the sector was highlighted. Construction innovation research is currently sporadic, and the differentiation within innovative practices between large and small organisations is not clear. This difference must be understood by policy makers, corporate leaders, futurists, scholars and all those involved with and within the construction industry. Thus, the aim of this book is to contribute to this underdeveloped area of innovation in small construction firms by offering a conceptually organised synthesis of relevant literature and case study material.

Chapter 2

Innovation demystified

Introduction

Construction firms are being increasingly challenged to successfully innovate in order to satisfy the aspirations and needs of society and clients, while at the same time improving their competitiveness in dynamic and abrasive markets. The Egan report laments that 'too many of the industry's clients are dissatisfied with its overall performance' (Egan, 1998: Paragraph 3), and proposes that the necessary service or product improvement and company profitability can be realised through innovations to enhance leadership, customer focus, integrated processes and teams, quality and commitment to people (Egan, 1998: Paragraph 17). The substantial contribution that small construction firms make to the output of the industry, as identified in Chapter 1, signifies the importance for this body of firms to improve their innovation performance if the performance of the industry as a whole is to move forward.

This chapter asks 'What is innovation?' The literature on innovation in small construction firms is synthesised and structured around a generic model to provide a holistic picture of current knowledge. Significant gaps in the understanding and practice of innovation in small construction firms are identified which severely hamper understanding of the myriad complex and systemically interactive issues embodied within the theory and practice of innovation. The gaps identified form the basis for a number of important questions that the authors propose form an integrating agenda for research that constitutes the subsequent chapters in this book.

Innovation in construction

The *Concise Oxford Dictionary* (1990: 610) describes the verb *innovate* as 'bring in new methods, ideas, etc: make changes' and innovation and innovator as 'make new, alter'. The synonyms listed against innovation in the *Oxford Thesaurus* (1991: 223) are 'novelty; invention; modernization; alteration'.

Innovation is an overused word, especially in construction. Design, invention and innovation are frequently confused, not least because those involved in construction come from a wide background, each with their own subtle interpretations of the term innovation. In architectural literature, the word innovation tends to be used to describe either the design approach of the architect or the appearance of the finished building; hence, the design of the building is 'innovative' or the designers have worked in a manner regarded as 'innovative' by their peers (Emmitt, 2002). From a contractor's perspective, innovation is more commonly associated with new materials or construction techniques.

Freeman (1989) defines innovation as the actual use of a non-trivial change and improvement in a process, product or system that is novel to the institution developing the change. Bowley (1960) divided studies of innovation in the construction industry into two main groups: those that change the product, and those that affect costs and availabilities. Bowley (1960) classified innovations in a range from those that result in new products to the consumer, to innovations that led to products that were, from the viewpoint of the consumer, no different from existing products but a perfect substitute. In short, there are innumerable ways of working out classifications of innovations, and the advantage of one over another depends on the particular purpose of the study. Other authors have used different terminology. Slaughter (1998) uses five types of innovation when looking at the implementation of construction innovations: incremental, architectural, modular, system and radical. The diverse use of the term 'innovation' in construction is not particularly clear, but it does emphasise an important issue – a clear definition is needed.

Freeman's (1989) earlier definition of innovation in terms of a process, product or system is supported by many other authors, including Parker (1978), Davies (1979), Utterback (1994) and Slaughter (1998). Further, Bradley (1989) goes on to make a distinction between the initial idea (invention) and the innovation process, which covers all stages of product development up to, and including, its launch or implementation (see Figure 2.1). In the manufacturing industry, the decision to market the product to potential adopters is the start of the diffusion process (Chisnell, 1995).

In terms of the construction industry, architects may consider products that are new to the market and also products that have been on the market for many years, but which they have only just become aware of because they are faced with an unfamiliar building type or usual detail. Thus, innovation also encapsulates the newness of the innovation, rather than the length of time it has been on the market.

Moreover, the innovation research field in the construction firm context is thus still very much in its embryonic stage. Innovation theory and practice are being drawn from established bodies of innovation knowledge predominately based on other industries (for example, see Barrett and Sexton, 1999), but they have not been sufficiently envisioned, embedded and evaluated in a construction context to form a robust body of construction innovation knowledge in its own right. We agree with the observation that '[construction] project-based, service-enhanced forms of enterprise are inadequately addressed in the innovation literature' (Gann and Salter, 2000: 955). This observation is extended further by commenting that to our knowledge the construction innovation literature often emphasises construction firms of large size, and that innovation in small firms has been generally ignored. As we saw in Chapter 1, considerable evidence indicates that there is a significant

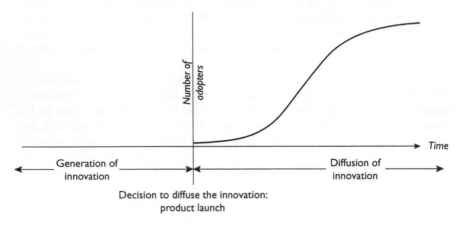

Figure 2.1   The generation and diffusion of innovation

difference in the innovation capability and output of small firms compared to large firms (for example, see Mansfield et al., 1971; Rothwell, 1989; Nooteboom, 1994; Rothwell and Dodgson, 1994). This difference must be understood, and underpin policy and corporate guidance.

Thus, the following offers a generic innovation model that aims to demystify what is meant by innovation in small construction firms. In doing so, it collates literature from the general innovation context and expresses it in such a way that is relevant for the construction sector.

Generic innovation model

The generic innovation model shown in Figure 2.2 argues that successful innovation *outcomes* are achieved through an appropriate innovation *focus* which is responsive to *contextual factors*, realised by appropriate organisational *capabilities* and channelled through effective and efficient innovation *processes* (Sexton and Barrett, 2003a). The model is similar to other 'generic' innovation models (for example, see Laudau and Rosenberg, 1986), in that it does not intend to capture the full social and economic dimensions of innovation (for example, see Rosenberg, 1982; Bijker et al., 1987). Rather, the model is offered primarily as a way of structuring this book, although the discussion herein will provide support for its general validity as a generic framework.

Focus and outcome of innovation

In the general innovation literature, a substantial part of the relevant debate on the focus of innovation concentrates on large firm innovation (for example, see

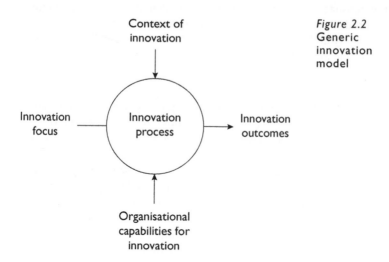

Context of
innovation

Innovation
focus

Innovation
process

Innovation
outcomes

Organisational
capabilities for
innovation

Figure 2.2
Generic
innovation
model

Woodcock et al., 2000: 214), and tends to cluster its definitional debate on innovation as being something new which is implemented by a firm in some way. Innovation is defined as 'generation, acceptance and implementation of new ideas, processes, products or services' (Thompson, 1965: 36) or the 'successful implementation of creative ideas within an organization' (Amabile et al., 1996: 25). The construction literature is consistent with the general literature, with, for example, innovation being defined 'as the actual use of a nontrivial change and improvement in a process, product, or system that is novel to the institution developing the change' (Slaughter, 1998: 226) or 'the process of bringing in new methods and ideas or making changes' (Atkin and Pothecary, 1994: 55).

What is interesting is that such definitions of innovation are 'value' neutral – namely, they do not explicitly state that innovation should add value in some way to the firm if it is to be deemed successful. This value neutrality highlights a dominant assumption in the literature; namely, that all innovation is beneficial. Kimberly (1981: 84–85) brings attention to this by noting that 'innovation tends to be viewed in unreflective positive terms . . . [and that] for the most part, researchers have assumed that innovation is good.' This assumption hampers an appreciation that innovation is associated with uncertainty and the risk of failure. Capaldo et al. (1997), for example, stress that innovation does not lead mechanically to improved performance – on the contrary, the decision to innovate may even strongly jeopardise the firm. The risk of such jeopardy leads to the 'innovator's dilemma' (Christensen, 1997): under which conditions firms should stick to what they already do and in which situation they should initiate innovation activity.

In the general literature there are hardening pockets calling for term innovation to accommodate the explicit benefit which must flow if innovation is to be considered successful. This is emphasised in the observation that innovation consists of the generation of a new idea and its implementation into a new product, process, or service, leading to the dynamic growth of the national economy and the increase of employment as well as the creation of pure profit for the innovative business enterprise.

(Urabe, 1988: 3)

Similarly, in the construction literature, for example, it is suggested that innovation is 'the act of introducing and using new ideas, technologies, products and/or processes aimed at solving problems, viewing things differently, improving efficiency and effectiveness, or enhancing standards of living' (CERF, 2000: 3).

In the general innovation literature, innovation is portrayed as having a number of roles or outcomes: the renewal and enlargement of product and service ranges and their associated markets; new methods of production, supply and distribution; and new organisational and work forms and practices (European Commission, 1995). In the construction literature, Thomas and Bone (2000: 67) identify three key areas for innovation activity which 'can deliver significantly improved quality and value': supply chain management and partnering, value and risk management, and technical innovation. Research into innovation in micro construction firms reported the following outcomes: use of computerised accounts and wages software, adoption of mobile telephones, adoption of cordless power tools, and adoption of computer-aided design (CAD) (Sexton et al., 2006).

In summary, within the general and construction literature there appears to be an ongoing shift from viewing innovation as an 'end' in itself, to innovation being a 'means' to achieve sustainable competitiveness. This debate, however, very much concentrates on large, non-construction firms, with little light being shed on the motivation for, and focus of, innovation in small construction firms. From this brief survey of the strategic focus and outcomes of innovation, the following issues appear significant:

- What is the general motivation for small construction firms to innovate?
- What generic strategic focus for innovation or definition of innovation does this motivation generate?
- What are common innovation outcomes in small construction firms?

Context of innovation

The necessary vision, legitimisation and investment for innovation are a product of shifting and intertwining forces both external and internal to the firm. This distinction between external and internal forces influencing firms' innovation trajectories underpins the two main schools of thought on what drives innovation: the market-based view of innovation, and the resource-based view of innovation.

The traditional market-based view of innovation argues that firms adapt or orientate themselves through innovation to optimally exploit changing market conditions. The general argument offered is that market conditions provide the context or initial conditions that either facilitate or constrain the direction and quantity of firm innovation activity (for example, see Porter 1980; Slater and Narver, 1994). Three strands of the literature investigating the context of innovation will be discussed here. First, there is the argument that the project-based nature of the construction industry is a significant barrier to innovation. Innovation often takes the form of pragmatic problem-solving on site which could not have been reasonably predicted before the project started. For such 'problem-solving' to become true innovation, 'the solutions reached for the particular problem faced on the project must be learned, codified and applied to future projects' (Winch, 1998: 273). Given the temporary nature of construction project teams and the short-term relationships between organisations that have come to be the norm, transfer of innovations from project to project and firm to firm has been extremely difficult (Construction Productivity Network (CPN), 1997).

Second, there is the potentially adverse effect on innovation of the structurally fragmented nature of the UK construction industry. The Egan report recognised, for example,

> that the fragmentation of the UK construction industry inhibits performance improvement . . . it is striking to note the number of small firms – there are some 163,000 construction firms listed on the Department of the Environment, Transport and the Regions' (DETR) statistical register, most employing fewer than eight people.
>
> (DETR, 1998: Chapter 1 Paragraph 8)

Case study research of innovation in SMEs has added support to this observation, with it being concluded that the structural characteristics of the UK construction industry restrict large-scale innovation and technology transfer, and that the capacity of SMEs to innovate is limited by their general inability to form long-term relationships with other firms (Miozzo and Ivory, 1998).

The third relevant strand from the market conditions and innovation debate is that information from the environment is presented in the form of 'precipitating events' that stimulate or hinder innovation activities (Zahra, 1991). Innovative firms tend to have the organisational aptitude and capability to be sensitive to these precipitating events and perceive them as imperatives or opportunities for pursuing innovation activity; conversely, non-innovative firms tend not to notice or act upon these activities (Miller and Friesen, 1984). Tidd et al. (1997: 14), for example, stress that innovative firms 'have to scan and search their environments (internal and external) to pick up and process signals about potential innovation . . . [these signals] represent the bundle of stimuli to which the organization must respond.' Julien (1996) consolidates this argument by asserting that

> firms cannot innovate if they cut themselves off or do not maintain their innovative contacts, if their information sources dry up and are not replenished,

if their networks decline or if information quality diminishes . . . and if they do not innovate, they decline and eventually disappear.

(Julien, 1996: 5)

In the construction literature Toole (1998), for example, concluded from a survey of one hundred SMEs that successful innovation within firms was fuelled to a significant degree by their ability to tap into many trusted sources of information, such as from other contractors, subcontractors and in-house expertise.

The recognition of a potential market-driven opportunity or imperative is a necessary condition for innovation, but not a sufficient one (Dosi, 1984). The adaptation and orientation to market conditions requires firms to choose appropriate strategies that are adequately resourced and implemented (for example, see Snow and Hrebiniak, 1980). The resource-based view of innovation focuses on firms' resources to understand their business strategies and to provide direction for, among other issues, innovation (for example, see Prahalad and Hamel, 1990; Andreu and Ciborra, 1996; Grant, 1997). The basic proposition here is that the market-driven orientation does not provide a secure foundation for formulating innovation strategies for markets which are dynamic and volatile; rather, firms' own resources provide a much more stable basis on which to develop its innovation activity, and to shape its markets, to a limited extent, in its own image. These resources may be physical, human, technological or reputational (Hadjimanolis, 2000: 264). The argument in the literature is that innovative small firms are those which can sense and act upon *internal* 'precipitating events' to create and develop unique resources or configurations of resources that serve as the foundations for successful streams of innovation.

Maijoor and van Witteloostuijn (1996: 549) suggest that 'the resource-based view of the firm seeks to bridge the gap between theories of internal organizational capabilities on the one hand and external competitive strategy theories on the other hand.' Indeed, research seeking to synthesise these two perspectives has suggested that 'while firms' resource endowments may determine strategy success, strategy choice is . . . restricted by market structure' (Hewitt-Dundas and Roper, 2000: 1). In the literature about innovation in construction SMEs, evidence to support this balancing of market conditions and resource endowments was reported in the observation that 'the sifting of possible [innovation] options was rigorous, with SMEs being close enough to both their markets and their capabilities to instinctively know what will work, and what will not' (Sexton et al., 2006: 20).

The literature on the market-based and resource-based views of innovation can be gainfully linked, as shown in Figure 2.3, by extending the argument that innovating firms are those which can sense and act upon 'precipitating events' in both external market conditions and internal resource conditions in an appropriately balanced and integrated fashion. This balancing of market-based and resource-based views is consistent with the quasi-evolutionary coupling model of innovation which integrates science-push and demand-pull (for example, see Rothwell and Zegveld, 1985; Schot, 1992; Steinmueller, 2000) along innovation trajectories (for example, see Sahal, 1981; Dosi, 1982). The 'precipitating event' perspective is very much embodied in the observation that

> innovation is fostered by information gathered from new connections; from insights gained by journeys into other disciplines or places; from active, collegial networks and fluid, open boundaries. Innovation arises from ongoing circles of exchange, where information is not just accumulated or stored, but created.
>
> (Wheatley, 1992: 113)

The optimal innovation balance of *market-based* and *resource-based* innovation is contingent upon the 'market-pull' and/or 'resource-push' implications of the prevailing *precipitating events*. The organisational challenge is to generate the balance required to provide an *appropriate innovation focus* which enhances overall performance – and in so doing, dynamically links the focus and context of innovation.

In summary, the general innovation literature provides abundant, if somewhat fragmented, material on the contextual factors of innovation: precipitating events, market-led innovation and resource-based innovation. The construction literature is far more limited in its treatment of these critical issues, particularly from a small firm perspective. From this examination of the context of innovation, the following issues appear significant:

• What are the key precipitating events, external and internal to small construction firms, which trigger innovation activity?

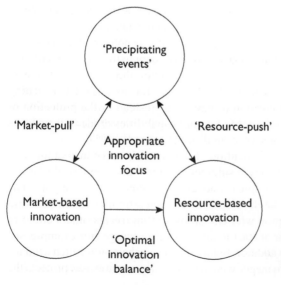

Figure 2.3
Synthesis of market-based and resource-based views of innovation

• What is the appropriate balance in emphasis between market-based innovation and resource-based innovation in small construction firms, and what conditions dictate this balance?

Organisational capabilities for innovation

Organisational capabilities for innovation are defined as 'the comprehensive set of characteristics of an organization that facilitate and support innovation strategies' (Burgelman et al., 1996: 8). This view of organisational capability is located within the dynamic capability literature which argues that appropriate capability development and exploitation geared toward effecting strategic change over time (Teece et al., 1997; Winter, 2000) give firms the foundation to generate multiple sources of sustainable competitive advantage (Barney, 1991). For our purposes we shall categorise capabilities into two distinct but complementary bundles: cognitive (or thought) capabilities, and organisational (or action) capabilities. Cognitive capabilities focus on the ability of individuals to innovate or be receptive to innovation; indeed, the foundation of innovation is ideas, and it is people who 'develop, carry, react to, and modify ideas' (Van de Ven, 1986: 592). Prerequisites to this flow of ideas are that there must be an initial cognitive trigger or felt need to innovate and the necessary power to progress these ideas. Taking the cognitive trigger first, individuals need to possess both the ability to organise and manage steady state activities for efficiency and reliability while still retaining a capability to identify key situations, both internal and external to the firm (for example, see Walsh and Ungson, 1991) where innovation is demanded in order to ensure effectiveness and responsiveness. In short, individuals need to be adept at 'switching cognitive gears', as illustrated in Figure 2.4 (Louis and Sutton, 1991).

In Figure 2.4 'automatic mode' equates to steady-state activities and 'conscious mode' to active problem-solving and innovation. It is stressed that the real problem is knowing *when* to switch from one to the other. This challenge is described as the 'management of attention' (Van de Ven, 1986). It was noted that management of attention is difficult because individuals gradually adapt to the environment such that their awareness of need deteriorates and their action thresholds reach a level where only crisis can stimulate action. The challenge for organisations is getting people to pay attention to the creation of new ideas instead of the protection of existing practices. In other words, not to let core capabilities erode in their focus and currency over time to become core rigidities (Leonard, 1995).

The issue of 'management of attention' feeds into the second prerequisite, namely, that the creation of ideas is not sufficient for innovation; among the issues, the idea must have adequate political and change management support. The literature often suggests that the development of a specific innovation in a firm requires an innovation champion who envisions and motivates others either to positively buy into the idea, or at least to allow it safe passage (for example, see Howell and Higgins, 1990). In addition, such innovation champions often need the benefit of a sponsor, a senior manager who symbolically nurtures and protects the

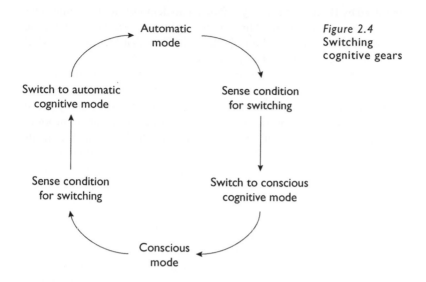

Figure 2.4
Switching
cognitive gears

innovation from political forces within the organisation who are hostile to the innovation (for example, see Maidique, 1980). The securing of a sponsor is argued to be significantly influenced by the ability of senior management to recognise the potential of a proposed innovation. This ability is argued to be a function of its managerial logic or view of the world, which in turn, depends on management experiences, organisational logic and industry logic (for example, see Nelson and Winter, 1982; Spender, 1989; Finkelstein and Hambrick, 1990). Evidence has been presented, for example, that many owners of small firms have a logic that is geared towards independence and autonomy rather than profits or growth (for example, see Bolton, 1971; Storey, 1986; Stanworth and Gray, 1991; Blatt, 1993; Gray, 1998). It is thus argued in the literature that the personality of these people has a significant influence on the innovative performance of small firms (for example, Miller and Toulouse, 1986; Dodgson and Rothwell, 1991; Rothwell, 1991).

Cognitive capabilities complement and interact with organisational capabilities, which both support and translate cognitive intent into organisational action. Organisational capabilities can be usefully grouped into functional and integrative capabilities (for example, see Verona, 1999). Functional capabilities are viewed as those which encourage a firm to deepen its knowledge base, while integrative capabilities broaden the knowledge base by capturing, blending and disseminating otherwise disparate knowledge across the company. The common currency which fuels both functional and integrative capabilities is knowledge. Knowledge contains both explicit (codified) and tacit (uncodified) elements (Polanyi, 1967), and people gain their knowledge, both explicit and tacit, through experiential learning (Kolb, 1984; Nonaka and Takeuchi, 1995). Research by Chaston et al. (1999) asserts that such learning is an antecedent of the organisational capability to innovate. This finding is endorsed by Barnett and Storey (2000), who demonstrated the important of such learning, concluding that small firms that are recognised as innovative have a proactive approach to learning, taking the holistic view that learning is a vital part of their long-term evolution and competitiveness.

The supportive space to develop and combine learning and innovation is linked to another central issue in the innovation debate: whether organisational slack promotes or hinders innovation. Organisational slack is defined as the 'pool of resources in an organization that is in excess of the minimum necessary to produce a given level of organizational output' (Nohria and Gulati, 1996: 1246). Proponents of slack resources argue that it facilitates innovation in firms by allowing risk-taking and experimentation that would not be supported in a more resource-constrained firm; similarly, it also releases managerial time and energy that would otherwise be focused on short-term operational performance issues rather than long-term strategic innovation activity (Cyert and March, 1963; Mohr, 1969). Opponents of organisational slack argue that it encourages unfocused, wasteful innovation activity which, in some cases, blinds firms to changes which are needed to meet external demands (for example, see Thompson, 1967; Jenson, 1986; Yasai-Ardekani, 1986). These two opposing schools of thought are being usefully synthesized through a contingency theory lens. Nohria and Gulati (1996) propose an inverse U-shaped relationship between slack and innovation in organisations, with both too much and too little slack being detrimental to innovation. The need for appropriate organisational slack very much underpins observations by Mason et al. (1996) for example, that there are two principal barriers to growth in small firms, namely a shortage of strategic skills and a limited range of mechanisms to stimulate business experimentation. Drawing the various aspects of capability together, research into innovation in small manufacturing-based firms reported that the

> accumulation and development of resources and capabilities are the relatively most important influential factors for innovativeness . . . managerial skills and capabilities, internal technological resources . . . and capabilities explain to a considerable extent the differences in innovation behaviour of small firms.
> (Hadjimanolis, 2000: 278)

From the construction literature, this view is supported for large construction firms, with it being reported that for successful innovation 'much depends on the unique needs and capabilities of each firm . . . key generic issues that emerged as being important were: climate, direction, commitment and knowledge management' (Barrett and Sexton, 1998: 23). In much the same vein, Tatum (1989) stresses the importance of creating an appropriate climate and capability pool for successful innovation which nurtures longer strategic horizons, risk tolerance and management, vertical integration of decision-making processes, flexible organisational structures, and proper matching of personnel to roles. The literature reviewed, however, did not extend their enquiry to the organisational capabilities for innovation in small construction firms. From this discussion of organisational capabilities for innovation, the following issues appear significant and in need of better understanding:

- What are the key cognitive and organisational capabilities for innovation in small construction firms?
- How are these capabilities developed and used in innovation activity?

Process of innovation

A distinction is often made between the content of innovation and the process of innovation. It should be noted, however, that process and content are mutually constitutive: the content of innovation is shaped by the process which generated it. There are two principal clusters of thought in the general innovation literature concerning the process of innovation in firms: the rational school and the behavioural school. The rational school of the innovation process is the most dominant, and considers innovation as multi-stage and linear in fashion. In the general literature the innovation process, for example, is seen as essentially 'organized, systematic, rational work' (Drucker, 1986: 40), comprising such stages as recognition, invention, development, implementation and diffusion (Maidique, 1980). Similar thinking underpins the construction literature's view of innovation process, with

> the process of introducing and commercializing innovations consist[ing] of several stages . . . an idea is conceptualized and defined; the idea is refined through research, testing and development to create an innovative process, product, or technology; the innovation is tested and demonstrated in laboratory and real-world settings; the innovation is further refined using data generated from testing and demonstration; the innovation is introduced to the marketplace for widespread application and use.
>
> (CERF, 2000: 3)

The rational orientated conceptualisations of innovation processes are criticised widely as they do not accurately portray the process of movement, interaction, and feedback of knowledge and resources within uncertain and dynamic environments (for example, see OECD, 1991). This challenge has attracted two responses. First, renewed efforts to firm up the rational process. The generic design and construction Process Protocol, for example, is based on the 'consistent application of the Phase Review Process irrespective of the project in hand. This together with the adoption of a standard approach to performance measurement, evaluation and control, will facilitate the process of continual improvement in design and construction' (Cooper et al., 1998: 2–21). Similarly, Motawa et al. (1999: 180) argue that 'the characteristics of construction innovation emphasise that traditional planning processes need to be developed to support more effectively the implementation progress of innovative projects.'

The second response has focused on endeavours to more adequately couple the innovation process with organisational reality. This emphasis has led to behavioural perspectives which argue in the general innovation literature, for example, that innovation is essentially 'controlled chaos' (Quinn, 1985), and that 'the innovation journey is a nonlinear cycle of divergent and convergent activities that may repeat over time and at different organizational levels if resources are obtained to renew the cycle' (Van de Ven et al., 1999: 16). This behavioural approach certainly flavours research findings into the briefing process which advocate solution areas focused around empowering the client, managing the project dynamics, appropriate user involvement, appropriate visualisation techniques and appropriate team building – rather than slavish adherence to linear briefing processes such as the Royal Institute of British Architects (RIBA) Plan of Work (Barrett and Stanley, 1999). Generally, however, the review of the construction innovation literature revealed scarce consideration of behavioural approaches in construction firms; and, in particular, small construction firms. From this discussion of the process of innovation, the following issue appears significant:

- Are the processes of innovation in small construction firms rational and/or behavioural in nature?

Summary

This chapter explored the various definitions of the term 'innovation'. Its use is particularly varied, not least in the construction industry as those involved come from a wide and diverse background, each with their own subtle interpretations of the word. It concludes by defining that successful innovation is 'the effective generation and implementation of a new idea, which enhances overall organisational performance'. Further, an idea can be classified as a process, product or system, and innovation constitutes the newness to an organisation as opposed to the length of time it has been on the market.

This chapter also explored general innovation literature and construction specific literature pertaining to innovation in small construction firms. The discussion was structured around the generic innovation model set out in Figure 2.1: innovation focus and outcomes; the context of innovation; organisational capabilities for innovation; and the process of innovation. The literature on innovation in small construction firms offers valuable fragments of insight, but the body of literature as a whole is characterised by a lack of clear, holistic direction. Significant gaps in the understanding and practice of innovation in small construction literature are identified which severely hamper understanding of the myriad complex and systemically interactive issues embodied within the theory and practice of innovation. Pieces of the research jigsaw are well developed, while other pieces are not – and all the pieces lack the overall picture to make sense of where the individual

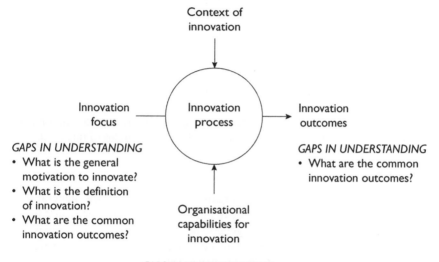

GAPS IN UNDERSTANDING
• What are the key precipitating events which
  trigger innovation activity?
• What is the appropriate emphasis between
  market-based and resource-based
  innovation?
• How do small construction firms sense and
  act upon the information generated from
  precipitating events?

Context of
innovation

Innovation
focus

Innovation
process

Innovation
outcomes

GAPS IN UNDERSTANDING
• What is the general
  motivation to innovate?
• What is the definition
  of innovation?
• What are the common
  innovation outcomes?

Organisational
capabilities for
innovation

GAPS IN UNDERSTANDING
• What are the common
  innovation outcomes?

GAPS IN UNDERSTANDING
• What are the key cognitive and
  organisational capabilities for
  innovation?
• How are capabilities developed
  and used in innovation activity?

Figure 2.5   Summary of key research questions

pieces fit into the broader portrait. The gaps identified are the basis for a number
of important questions which are proposed as an integrating agenda for future
research. These are summarised in Figure 2.5.

The research questions identified potentially enable otherwise disparate research
activity to be brought together synergistically within a generic innovation model.
This model has the potential to provide a design tool to create balanced innovation
in small construction firms, research programmes, projects and events. Each aspect
will be explored within the remainder of this book.

Chapter 3

Case study investigation

Introduction

It has so far been observed that although the mainstream innovation research field is substantial, within the context of construction, it is minimal. This is particularly pertinent for small construction firms. To investigate innovation in the small construction firm, the case study research approach has been adopted. This chapter introduces the cases which will be drawn upon in the remainder of this book.

Research aim

Chapter 2 furnished a conceptually organised synthesis of the general and construction specific literature pertaining to innovation in small construction firms. The findings were structured around a generic innovation model which argues that successful innovation *outcomes* are achieved through an appropriate innovation *focus* which is responsive to *contextual factors*, realised by appropriate organisational capabilities and channelled through effective and innovative processes (see Figure 2.2, p. 13). The model is offered primarily as a way of structuring the study, although the literature review provided support for its general validity.

The generic innovation model shown in Figure 2.2 is similar to other 'generic' innovation models (for example, see Laudau and Rosenberg, 1986), in that it does not intend to capture the full social and economic dimensions of innovation (for example, see Rosenberg, 1982; Bijker et al., 1987). Rather, the model is offered primarily as a way of structuring the investigation of innovation in small construction firms in this book.

The synthesis identified a number of research questions structured around the generic innovation model. These are listed in Table 3.1 and were designed as key questions by predicating that innovation in small construction firms does not occur in a vacuum.

The case studies

Bearing in mind the need to distinguish between large and small sized firms, specifically in relation to innovation, the emphasis of this study is the generation

Table 3.1 Research questions

Generic innovation model: focus and outcome of innovation

- What generic strategic focus for innovation or definition of innovation does this motivation generate?
- What is the general motivation for small construction firms to innovate?
- What are common innovation outcomes in small construction firms?

Generic innovation model: organisational capabilities for innovation

---

- What are the key capabilities for innovation in small construction firms?
- How are these capabilities developed and used in innovation activity?

Generic innovation model: context of innovation

- What are the key events external and internal to small construction firms which trigger innovation activity?
- What is the appropriate emphasis between market-based innovation and resource-based innovation in small construction firms, and what conditions dictate this emphasis?

Generic innovation model: process of innovation

- Are the processes of innovation in small construction firms rational and/or behavioural in nature?

---

of new theory rather than trying to apply theory that was relevant to large firms. To support this, variety in the data was needed. It was therefore decided that a number of small firms would be invited to take part in the project and that the project would look at two areas of the construction supply chain, namely, construction consultants and construction contractors (Sexton and Barrett, 2003b).

Eight small firms were initially selected to take part in the project but, soon after the project commenced, one dropped out due to work commitments. The seven firms that took part are all from the North West of England. Three of these firms are contractors, the other firms are consultants, namely firms of architects, building services engineers, building and quantity surveyors, and quantity surveyors. Table 3.2 summarises some of the key features of the participating firms. All of the contracting firms are limited companies. Three of the consulting firms are partnerships, with the fourth being a limited company. The firms are denoted by letters to give the collaborating firms as much anonymity as possible.

We acknowledge that the sample set of firms taking part in the project may not be representative of the industry as a whole. (Over 200 firms were contacted in order to secure eight firms who wanted to be involved in the research project and which were the right size.) We accept, therefore, that those firms that have taken part have possibly done so as they are more interested than others in progressing

Table 3.2 Key characteristics of participating small construction firms

Company	Est.	No. owners	Turnover (1999)	Staff no.	Principal business focus	Main fields of expertise or clients
Consultant A	1997	4	£1.25m	26	Architectural design and project management	All major commercial sectors (retail, office, financial, art and leisure fields)
Consultant B	Mgmt buy out 1998	2	£0.44m	11	Quantity surveying and construction cost consultancy	Public and private sector clients
Consultant C	1981	3	Information not provided	20	Survey, design and building contract administration	Mostly public sector, with some commercial and industrial clients
Consultant D	1990	4	£0.5m	20	Building services engineering	Mechanical, electrical and public health services (including data and telecom systems) for commercial, health care, education and industrial projects
Contractor A	1981	2	£2m	25	General building contracting	Local authority and house association
Contractor B	1996	2	£3.2m	15	Management of building and engineering contracts	Blue-chip clients in petroleum, food retailing, commerce, industry and civil engineering
Contractor C	1985	2	£2.12m	25	Renovation and refurbishment; quality fit-out contractor, project management and subcontractor coordination	Mainly commercial and some industrial blue-chip clients and local authorities

their businesses through innovation, and therefore are potentially atypical of the broader population of small construction projects.

A number of research techniques were utilised in each case, including interviews with key personnel (to understand the nature of innovation), problem-solving workshops (to provide the momentum and interaction necessary to provide a highly stimulating context for investigating innovative activity) and action research steps (to investigate particular aspects of innovation). The results were triangulated to provide generalisable results.

Summary

This chapter presented the research methodology followed in order to investigate the cause and impact of innovation in small construction companies. A number of case studies were introduced and the process of investigation was described. Multiple perspectives from the small firms were sought to add to the richness to the data, and a cross-case method, which triangulated the data, was needed to provide consensus and generic themes.

The subsequent chapters will contextualise the outlined research issues identified within the relevant general and construction-specific innovation literature, and use the findings of the study to illustrate how they relate to small construction organisations. It will promote the benefits of innovation by stimulating the innovation capability within the following key areas:

- the focus and outcome of innovation
- the organisational capabilities for innovation
- the context of innovation
- the process of innovation.

Chapter 4

Focus and outcome of innovation

Introduction

Chapter 2 argued that not all innovation per se is beneficial; rather, *appropriate* innovation is beneficial. Thus, it is important to define the focus and expected outcome of the innovation at the outset. This chapter provides empirical evidence from case study research to substantiate this. The chapter will address the relevant questions detailed in Chapter 2:

- What generic strategic focus for innovation or definition of innovation does this motivation generate?
- What is the general motivation for small construction firms to innovate?
- What are common innovation outcomes in small construction firms?

What generic strategic focus for innovation or definition of innovation does this motivation generate?

Successful innovation is defined as 'the effective generation and implementation of a new idea, which enhances overall organisational performance'. This definition was developed originally by (and for) large construction firms (Barrett and Sexton, 1998), but was considered by the participating practitioners to be sufficiently inclusive to accurately define innovation in small construction firms. Applying this definition to small construction firms, the following assumptions are emphasised and illustrated:

- *Idea* – ideas are taken to mean the starting point for innovation (for example, see Thompson, 1965). Ideas can be administrative in nature (for example, the organisational restructure and process changes to support partnering carried out by Contractor B) and technical in character (for example, the computerisation of quantity surveying computation and report generating tasks by Contractor C).
- *New* – not all ideas are recognised as innovations and it is accepted that newness is a key distinguishing feature (for example, see Zaltman et al., 1973). The idea

has to be new only to a given firm, rather than new to the 'world'. The use of mobile telephones to improve site-based communication by Contractor A is new to that firm, even if it is a fairly well-established technology in other industries and firms. Further, the newness aspect differentiates innovation from change. All innovation implies change, but not all change involves innovation. For a contractor, for example, a change in a materials supplier is not necessarily an innovation, but a change in the relationship between the contractor and the supplier from a project-to-project open tender situation to a long-term 'partnering' type of relationship would constitute an innovation. The examples of innovation offered by the firms tended to be the adoption of established ideas or technologies and/or their incremental adaptation. This is consistent with the view in the literature that small to medium sized firms rarely introduce fundamentally new products to their industry (Storey and Sykes, 1996), but are more likely to be involved in making incremental changes based on generic technologies than engaged in more transformational changes (Rosenberg, 1992).

• *Effective generation and implementation* – innovation requires not only the generation of an idea (or transfer of a 'new' idea from outside the company), but also its successful implementation (for example, see Thompson, 1965). The implementation aspect differentiates innovation from invention (for example, see Monk, 1989).

Box 4.1 provides a mini case study which illustrates the various dimensions of innovation discussed above. Innovation must improve overall organisational performance, either individually, or collectively through the supply chain (for example, see Kimberly, 1981). Innovations that improve some isolated aspect at the expense of overall performance are undesirable. This thrust is clear in the assertion by Contractor B that the use of CD-ROM technology to improve the marketing activity would be a way to improve the firm's overall performance, not just the marketing performance.

Thus, the key implication of innovation is that not all innovation per se is beneficial; rather, appropriate innovation is beneficial.

What is the general motivation for small construction firms to innovate?

The general motivation to innovate within small construction firms is depicted from the case study findings as, first and foremost, to generate sufficient cash flow to survive in the short term. This imperative is epitomised by both contractors and consultants alike in the following observations:

In the business context, if we weren't severely affected by variations in workload, then we could sit back and think, but when you're focused on where the opportunities are, what's coming through the door, how many tenders have

Box 4.1 The improvement of Contractor A office–site communications through the use of mobile telephones

Idea

The starting point of the innovation was the perceived need to improve communication between the office and on-site staff. The managing director of the firm already had a mobile telephone, and thought this technology could be deployed throughout the firm.

New

The use of mobile telephones to improve office–site communications was certainly not new to the world, but was new to the firm

Effective generation and implementation

The idea of using mobile telephones was absorbed from outside the firm, and has been successfully implemented, with all relevant staff now having a mobile telephone.

Enhances overall performance

The use of mobile telephones has improved the supervision of site staff and, if there is a problem, the site staff can be in contact with the office personnel, the architect, the client or subcontractors. The more timely communication provided by mobile telephones also enables resources to be managed more efficiently: 'You might have five cubic metres of concrete being delivered today and you've had to organise four men to spread it when it gets on site. If one of them doesn't turn up, then the mobile phone comes into its own, because you can ring up another site and say "Look, you get over here".'

you got this month, what turnover is that producing, what's the potential return etc., then you don't have the luxury to sit back and contemplate how you might do something better . . . until businesses like ours are taken off the bread line, they're not going to have the thinking time to innovate.

(Contractor B)

If we are struggling in terms of profit or turnover, the first thing to go is any time spent working on new systems and innovating; you just do what you can, the best you can, with what you've got.

(Consultant C)

> Profit comes first, everything else comes second, because if you don't have profit you don't innovate because you can't afford it and you can't develop your practice and your team.
>
> (Consultant A)

The 'short-term survival' motivation limits innovation activity to one-off, project-specific 'problem-solving'. This position reflects both the uncertainty of income streams and the limited resource base of the firm, depriving them of the kind of market and resource buffer which large firms often use to smooth out the peaks and troughs of workload. These findings are certainly consistent with the literature, with Storey and Cressy (1995) demonstrating that the competitive environments in which small to medium sized firms tend to operate increase the risk of failure. The abrasive business environment conditions are compounded by research that identifies limits on capitalisation and lack of liquidity as being major threats to the survival of small to medium sized firms (for example, see de Koning and Snijders, 1992).

The instrumental role of market and resource buffers in innovation is also supported by the innovation literature (see Chapter 2), which argues that organisational slack (or excess resources) is required to support the risk-taking and experimentation intrinsic to innovation activity, and that there is an optimal level of slack, with both too much and too little slack being detrimental to innovation (for example, see Nohria and Gulati, 1996; Cheng and Kesner, 1997). The case study research findings indicate that small construction firms lack the required slack to innovate to any significant degree at the 'survival' stage, and that more slack is needed to enable such innovation activity to take place.

It is only after the ongoing challenge of survival has been achieved and some degree of stability is secured that attention appears to then be directed to 'developmental' innovation activity to increase profitability and/or growth – through such things as 'repeat business' and 'new business opportunities'. This need for stability before development can take place is evidenced by both contractors and consultants:

> Long-term stability would allow us to plan for the future, and perhaps have a formal development budget, and devote human resources to development.
>
> (Consultant C)

> From being small, we think big and therefore we have to be . . . at some point [we] have to take a quantum leap and it is when to do that and when you have the funds to do that.
>
> (Consultant A)

> I think it's more likely that once we get the sustainable level of throughput . . . then we will look at trying to refine and improve the service that we give . . . excellent results on projects rather than satisfactory or good.
>
> (Contractor B)

If we don't provide the service that clients want then we don't get repeat business, and growth becomes very difficult.

(Contractor C)

In summary, the case study results indicate that the motivation to innovate follows a fluid hierarchy of 'motivational needs' (see Figure 4.1):

- *Survival* – small construction firms, owing to the type of markets they operate in and their lack of organisational resources and slack, concentrate foremost on project-based innovation focusing on survival.
- *Stability* – it is only once survival has been confidently achieved that firms are sufficiently motivated to look towards consolidating and stabilising their market and/or resource position to ensure steady-state conditions over the medium term.
- *Development* – this stability provides the necessary motivation to exploit the prevailing stability and to develop and/or grow.

This type of hierarchy is consistent with arguments located in the stage theory literature. This literature describes stages 'as a configuration of organizational design variables representing a firm's response to the sets of dominant problems it faces at sequential times' (Kazanjian, 1988: 257).

Flavoured by this type of thinking, Churchill and Lewis (1983), for example, describe five stages through which small firms pass: existence, survival, success, take-off and resource maturity. The case study findings, however, depart from this literature, by emphasising that survival, stability and development stages are not rigidly linear in progression, but cyclical in response to dynamic market conditions. This dynamic and cyclical behaviour confirms that small construction firms remain

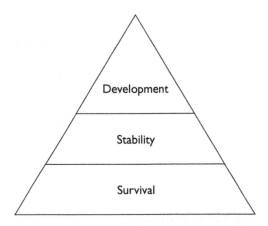

Figure 4.1 Hierarchy of motivational drivers for innovation

more open to their external environments compared to large firms owing to their comparative lack of market and resource buffers. Consultant C, for example, commented on the impact of the recession in the late 1980s:

> It just changed overnight; competition grew, fees were driven down as a percentage, and, because of the recession, the building-works costs had been driven down as well. Resulting in a double whammy: a smaller percentage of a smaller amount. We had to close the Bristol office, and reduce the size of the firm substantially . . . turnover has probably just grown back to what it was in the late 1980s.

Further, there is a tendency in the stage theory literature to assume that small firms strive for growth per se. The case study findings indicate that the motivation to innovate is not solely to grow, but can be directed at creating sustainable steady-state development. Consultant C, for example, stressed:

> In terms of things like expansion – opening offices and geographical expansion – we probably have very low expectations there. A partnership structure doesn't really lend itself to that. There's a limit to the amount of money you can make as a partnership and as a partner, because the more the turnover you make the more partners you need, so you're sharing more money out . . . I think the three partners here are all hitting or just hitting their fifties so maybe our strategy is a little bit old fashioned because we're looking more to retirement.

This is supported in O'Farrell and Hitchens' (1988) argument that:

> whether their [the small firm's] desire to remain small is a rationalization of their lack of capability and resources is not of crucial importance; they are content to stay small; and policy instruments designed to aid in the process of growth in small businesses are likely to have little or no impact upon such firms.
>
> (O'Farrell and Hitchens, 1988: 1369)

It should be noted that although the case study findings stress the limited amount of organisational slack to enable innovation activity, it should not be presumed that more slack necessarily means better, large-scale innovation. As observed in the literature (for example, see Nohria and Gulati, 1996; Cheng and Kesner, 1997), there will come a point where additional resources channelled into innovation activity will generate ever decreasing returns, and will, in effect, be a waste of precious finite resources.

The principal implications of the case study findings on the motivation for small construction firms to innovate are threefold. First, small construction firms are not always motivated to innovate; when in 'survival' posture, firms will generally want to limit their exposure to the costs and risks of innovation as much as possible. Second, the hierarchy of motivational drivers for innovation (survival, stability and development) are dynamic and cyclical, not a linear progression. Third, not all small firms want to grow indefinitely in size; firm size will stabilise at a level which is compatible with the owners' aspirations.

What are common innovation outcomes in small construction firms?

The findings provide numerous and diverse examples of innovation outcomes: client relationship development innovation, organisation and management innovation at firm and project levels, technological innovation, etc. The principal outcome of innovation activity can be usefully grouped into two areas: improving the effectiveness of the firm, i.e. making sure that the firm is doing the right activities, and improving the efficiency of the firm, i.e. making sure that the firm's activities are done well.

Summary

This chapter has discussed the focus and outcomes of innovation in SMEs. However, this is not sufficient in itself to bring about successful innovation; firms need both the organisational capability and an appropriate response to the inter-action environment to innovate. These issues are discussed in Chapter 5.

Chapter 5

Organisational capabilities for innovation

Introduction

As discussed in Chapter 4, SMEs articulating a particular innovation focus and the anticipated benefits flowing from this focus are not sufficient in themselves to bring about successful innovation. Firms need both the organisational capability and an appropriate response to the 'interaction environment' to successfully innovate. This chapter considers the key capabilities for innovation, and further, how these capabilities are developed and used.

What are the key capabilities for innovation in small construction firms?

The case study produced a model of the organisational factors critical to successful innovation (see Figure 5.1) which proved to be useful in both understanding and managing innovation activity, i.e. it is both an analytical and prescriptive model. The variables which make up the model are defined as follows:

- *Business strategy* is concerned with the overall purpose and longer term direction of the firm and its financial viability.
- *Market positioning* is the chosen (or emergent) orientation towards desired target markets for the purpose of achieving sustainable profitability.
- *Technology* is the machines, tools and work routines used to transform material and information inputs (for example, labour, raw materials, components, capital) into outputs (for example, products and services).
- *People* are viewed as possessing knowledge, skills and motivation to perform a variety of tasks required to do the work of the firm.
- *Organisation of work* involves the creation and coordination of project teams and commercial networks both within the firm and across its business partners.
- *Interaction environment* is that part of the business environment which firms can interact with and influence.
- *Given environment* is that part of the business environment which firms are influenced by, but which they cannot influence themselves.

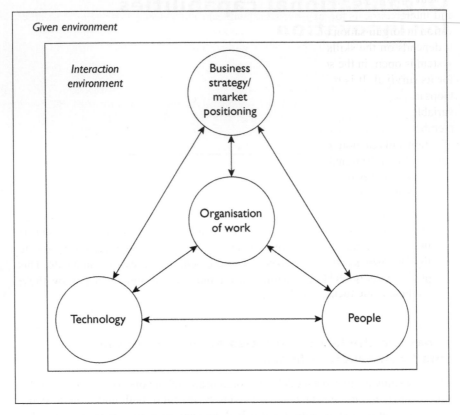

Figure 5.1 Organisational factors of innovation model

The model proposes that business strategy, market positioning, organisation of work, technology, and people are the key organisational variables in understanding and improving innovation in small construction firms. The model emphasises and embraces both the holistic and systemic dimensions of innovation. The creation, management and exploitation of innovation involves consideration of not only the *content* of a chosen innovation, but also the management of the *process* of innovation and the *context* in which it occurs. The model considers two aspects of context: the inner and outer contexts of the firm. The inner context refers to the business strategy, market positioning, organisation of work, technology and people. The outer context refers to the given and interaction business environments. The process of innovation refers to the actions, reactions and interactions of, and between, the various organisational variables in the outer and inner contexts.

The organisational model argues that for practitioners to understand and organise innovation successfully, a broad, outwards looking approach is needed. It is a system, in the sense that all of the variables, shown in Figure 5.1, are interdependent and interrelated, as indicated by the double-headed arrows. For example, innovation in 'organisation of work' shapes and is shaped by all of the other variables. It depends on the skills of the people and technology available to the firm. The system is open, in the sense that it depends on the external business environment for its survival. It is recursive, in the sense that it assumes patterns of feedback loops and sequences of interaction, which link and integrate the key organisational variables during innovation or change. It is dynamically complex, in the sense that it embodies multiple patterns of cause and effect over time; and changing patterns of action and reaction, relationships, meanings and rules. The holistic or systemic nature of organisational work, and thus the role of innovation within it, is captured by the assertion that:

> The crucial thing, to make everything work, to make it viable, is to provide something that your client wants, and suits all their requirements in terms of how much it's going to cost, what we are going to deliver, how we will deliver it, and a whole range of things. You've got to find something that we have to do. We have to find a fit between our requirements and their requirements, and then persuade them that it's beneficial to them as well as to us. And then we have to sit down and plan the method of delivery, and that involves all those things here; recruitment, training, learning, and organisation and structure, and setting that up with the right management, leadership. It's a matrix isn't it.
>
> (Consultant C)

The model is very much located in the contingency systems school of thought. Kast and Rosenzweig (1973) summarise the general shape and content of this thinking in the statement that:

> the contingency view of organizations and their management suggests that an organization is a system composed of subsystems and delineated by identifiable boundaries from its environmental suprasystem. The contingency view seeks to understand the interrelationships within and among subsystems as well as between the organization and its environment and to define patterns of relationships or configurations of variables. It emphasizes the multivariate nature of organizations and attempts to understand how organizations operate under varying conditions and in specific circumstances. Contingency views are ultimately directed toward suggesting organizational designs and managerial actions most appropriate for specific situations.
>
> (Kast and Rosenzweig, 1973: ix)

Each of the variables in the model are discussed more fully in the next section, starting from the business environment.

How are these capabilities developed and used in innovation activity?

The 'interaction' environment is defined as that part of the business environment which firms can interact with and influence. This includes existing and potentially new clients, consultants, competitors, contractors, subcontractors and suppliers. The case study project findings identify clients as being the principal agent in the interaction environment. Consultant D, for example, generated 50 per cent of its turnover from healthcare work and, by May 2000, had been advised that the healthcare sector was deciding whether to change their procurement practice because the government did not feel that the various National Health Service (NHS) Trusts had been particularly efficient at procuring new buildings. The proposed changes to the procurement regime are perceived by the firm as being potentially disastrous for its business. The firm is trying to mitigate this risk by talking to other health service trusts about the services they can provide, as well as diversifying into other areas such as call centres so that no one single market accounts for more than 30 per cent of its turnover.

In contrast to the interaction environment, the 'given' business environment is that part of the business environment which firms are influenced by, but which they cannot influence themselves. This includes things like interest rates, public-spending budgets and building regulations. Changes in the given environment are viewed as having potentially significant effects. Consultant C, for example, notes that:

> Fluctuations in the business cycle are the biggest problem. Whenever there is a lack of confidence in the economy or a downturn, everyone stops spending on building because they can do that without affecting their profits. People will build when they're making profits, and they will repair their building when they're making profits; but if they are not making profits, that's the first thing that stops.

The remoteness of some of the key factors in the given environment to the day-to-day operation of small construction firms is captured in Contractor B's observation that

> There's not a lot that you can do about the government – so you can take them out of it. It's an 'across the board' thing.

This view is echoed by Consultant A, who notes that responding to regulation requirements, though time consuming, is something 'which you always have to do'.

The differentiation of the business environment into given and interaction domains is consonant with the literature. Kotler (1980), for example, identifies four levels of the environment which surround a company: the *task environment*, which comprises the main stakeholders in performance of the firm's task, such as suppliers and consumers; the *competitive environment*, which consists of other firms which compete with the firm for customers and scarce resources; the *public environment*, which comprises institutions which regulate or impinge in some way on the activities of the company or the sector in which it operates, and the *macro environment*, which consists of the main social, legal, economic, political and technological forces which confront the company. The given environment developed in the case study project is taken to be synonymous with the public and macro environment, while the interaction environment is viewed as encompassing both the task environment and the competitive environment.

The given and interaction environments provide the outer context within which small construction firms operate, and which significantly determine the nature and scale of innovation within small construction firms. The focus for now will be the inner context (the business strategy, market positioning, organisation of work, technology and people variables), which both shapes and is shaped by the given and interaction environments, and is discussed below.

Business strategy and market positioning

The business strategy and market positioning variables in the model are defined as follows: business strategy is concerned with the overall purpose and longer term direction of the company and its financial viability, and market positioning is the chosen (or emergent) orientation towards desired target markets for the purpose of achieving sustainable profitability.

Business strategy

The case study findings indicate that small construction firms have business strategies which are 'soft focus' in nature. The term 'soft focus' denotes a business strategy which maps out a broad strategic aspiration, but that aspiration is not fleshed out in too great a detail in terms of what the firm wants to achieve, how it wants to achieve it, and when it wants to achieve it by. The soft focus provides both a cue and a vehicle for strategic decision-making and action, rather than a rigid goal and model. Contractor B, for example, articulates its business strategy in the following way:

> [we] want to grow turnover, and also not to be in a position where so and so who knew us from x, y or z drops an enquiry through the door and we take that, or we've got one bid going in and we've got to make sure it's a good one; we want to be in a position to pursue say this avenue or that avenue.

Similarly, Consultant D stresses that having some of strategy is essential because

> otherwise you do not really know where you are going. You may end up deviating off it, but at least you have something to aim for in the first place.

Business strategy is depicted by other firms in a more fluid, moment-by-moment fashion:

> Every so often, perhaps quarterly, the Associates and us, perhaps just go out for a pizza somewhere. We then say, 'Right, where are we going? What do we need to do? Any thoughts?'
>
> (Consultant B)

> Every month the directors evaluate the business through various means: cash flows and answers to what's going on . . . what's on board and in the pipeline.
>
> (Consultant A)

However, the nature and volatility of their workload tended to create reactive responses. This position is captured by Consultant B, who argues that despite the strategic rhetoric espoused in 'pizza' meetings, the reality is that

> our strategy is very much driven from the outside by clients. If they want something done quicker or in particular formats, we have to adapt. They're still driving the way the industry moves, as much as we try to guide things in a certain direction. I don't think we change the behaviour of the client from a strategic point of view Technically we might say that we've got this form of contract, or we recommend this contractor, but I think from the point of view of the client wanting a new building or upgrading an existing building, I don't think we have an awful lot of influence. That decision's usually been made by the time it gets to us . . . we tend to be more reactive than proactive. It is difficult for a practice our size to be proactive.

The tension between the aspiration of long-term business planning and the volatility of workload is stressed by contractors also, with Contractor A stating that they do not look beyond the length of their longest contract at any given time (at most twelve months) because: 'in our industry you can't, because you don't know what is going to happen. You get long established companies . . . going out of business.'

The apparently contradictory 'reactive' nature of strategic thinking resonates with the observation by Bracker and Pearson (1986) that small firms tend to focus on adaptation issues, while larger firms concentrate on integration issues. This reactive, adaptation orientation is considered to have a positive aspect by the small firms when compared to large construction firms:

> Responses within an organisation like ours tend to be much quicker than those in larger firms. It isn't a long-winded process, where policies are formalised, and written down, and information is disseminated by memos and letters and emails. People just meet on corners, saying, 'We're going to do this tomorrow.' If someone has an idea, they walk into another partner of the office and say,

'What about this', and they say 'Yes, go away and do it. Ring him up, get on with it.' In a small firm we can pick something up and run with it very quickly.

(Consultant C)

This perception endorses the literature which argues that small firms are often more agile and responsive than larger firms (for example, see Mansfield et al., 1971; Rothwell, 1989; Nooteboom, 1994; Rothwell and Dodgson, 1994).

In addition, the case study findings show that this reactive stance is exacerbated by the lack of managerial time and expertise which tends to create tactical responses to day-to-day opportunities and obstacles. Contractor C comments, for example, that although it is able to 'nail issues' quickly, the firm exhibits weakness in addressing longer-term issues as resources are committed to work-in-progress; and the 'managers, like many in construction firms, have neither the experience nor training to manage long-term investments.'

Similarly, Contractor B explains that the two directors of the firm have

come from such an operationally focused background . . . and that's where we've concentrated our efforts, to make sure that we're working efficiently in terms of what we do out on site – how we build. But in terms of the overview, and how the front end of the business works – getting new work – we had been totally underexposed in the past, so we've had to learn quite quickly how to do it.

Closely linked with the effect of managerial perception of business strategy is the substantial role that the owners have in influencing the business strategy of their firms. The owner of Contractor A, for example, comments on the power of his position in the observation that:

They probably know that we are going to do it . . . when you have 19,999 shares out of 20,000 you don't have resistance. That is the reason you have all the shares . . . It's just like anything else, it's mine.

This position was evident in Consultant B, with one of its two directors commenting that the other director effectively made the decisions, explaining:

it is nominally a joint thing, but he has more of a shareholding than me. At the end of the day someone has to be the leader and make the decisions.

Similarly, Consultant D observes that technical innovation often comes from the directors, in the sense that they give technical direction, and ask staff to consider alternative solutions.

This finding is consistent with the literature, which argues that the managerial logic or 'view of the world' exhibited by the principals of small firms have a considerable impact on envisioning and guiding strategy (for example, see Storey, 1986; Dodgson and Rothwell, 1991).

The financial viability constraints affecting the capacity and capability of small construction firms are epitomised by Consultant B, who stresses:

> small firms have a tight budget, so they don't have the people around to tackle a specific problem . . . the cost of innovation is the short-term human involvement, and then having committed the capital to physically spend, you need some human time to make it work. The three go together. The big one though is the cash one.

Together, the instrumental role of managerial time and expertise, and financial resources in this orientation, is consistent with Welsh and White (1981), who noted that the scarcity of resources, in addition to the knowledge, experience, perceptions and amount of time available to the principals united to produce a strategy formulation process which was distinctly different from that of larger companies.

Market positioning

In contrast to the 'soft focus' nature of the business strategy, the case study results suggest that small construction firms have a 'hard focus' on their market positioning, having in-depth knowledge about their target markets, and significant emphasis on developing and projecting a market orientation to those markets.

The depth of knowledge about their target markets is a function of two interactive factors: the level of repeat work, and the explicit effort to develop fruitful relationships with clients. The significant level of repeat work is reported by consultants and, although to a lesser degree, by contractors. Consultant A comments that it

> tends to have a rolling relationship with certain people, at least three or four agents. These relationships are developed like any other through trust and track record.

Similarly, Consultant C reports: 'about 75 per cent of our work comes from the same client every year . . . [and that] a lot of work, especially stock surveys, is due to referrals'.

For the contracting companies, the sources of work were more varied. Contractor C shared the same view as the consultants, confirming that it had long-term relationships with clients, and that it benefited from 'a lot of repeat business, via recommendation'. The firm, however, did add a note of caution, saying that: 'it's a cut-throat business; the big boys are ruthless and have no loyalties.' It is very much motivated by this 'ruthlessness' that Contractor C avoids working with large contractors.

The high level of repeat work and referrals is consistent with Hoxley (1993), who investigated the sources of work for building surveying practices, and concluded that some 60 per cent of work comes from existing clients, and that of the 40 per cent which comes from new clients, approximately 60 per cent comes from recommendations or referrals.

The contractors generally perceive that their sources of work are based on more transitory clients and open tendering arrangements. Contractor B noted that there is an 'over supply of people who do what we do'. This 'over supply' consideration leads to the type of response from Contactor A, who emphasises that they have a

> large client base because of the nature of the market; 'have to be in the right place at the right time or know the right person'; reputation is not an issue.

In summary, the consultants appeared to rely more on repeat work and work coming from referrals than the contractors. In addition, the procurement climate is such that contractors are more likely than consultants to have to tender on an open tender basis (rather than negotiation), regardless of whether the client has provided repeat work in the past or not.

The level of repeat business and referrals is created and nurtured by developing and projecting a market orientation to those markets. Consultant A, for example, positions itself in the retail and office markets, asserting: 'that's our strength and that is something we pride ourselves in and market.'

The importance of matching market opportunities to the advocated market positioning of the firm, rather than trying to take on work which was inconsistent with their market positioning, is evident in the observation by Contractor C that:

> I had an enquiry from a very good client who wanted me to design and build a church . . . We have a strong relationship with the client that brought it to us. I said to the lads, 'What do you think of this?' They looked at it, 'Lot of design input, got to be finished by August . . . No'. So we killed it there and then. We didn't waste time on it. We have to be selective, because I know we're not experienced with dealing with that sort of project. This other job is a variation on a theme; it's all stuff we're well capable of, and understand, and we run with all the time.

The small construction firms do not restrict themselves to one market, however; it is stressed that they need to operate in multiple markets to both capitalise on, and protect against, opportunities and threats which emerge over time in those markets. Consultant B, for example, comments:

> I do not think we can say that we are going to be specialists in hospitals' work or prisons' work or whatever it is and that we are going to do just legal cases or design and build. I do not think we can set out our stall like that. We need to try and keep some flexibility, and try and respond when things happen. Flexibility is an asset in terms of being able to respond to changes, but because of the size of our firm it can be difficult to be proactive and anticipate future changes.

Similarly, Consultant C explained:

> Obviously we try to plan a client base that is insulated from variations in funding. If, for example, all your work comes from Housing Associations and the government system of funding changes then you're in trouble; but if you've got half your work with Housing Associations and half your work with the commercial sector then you're a bit more insulated from those changes in funding. We're looking to find a range of clients in diverse sectors of the economy, who are affected differently by changes in the economy. That will, to some extent, insulate us from those changes. When I first started it used to be that you should do one-third public sector, one-third commercial, one-third industrial, for example, although that doesn't usually work because in a slump everyone goes down. That's the ideal situation, where you source your work from different client bases, so that you might have industrial work, commercial work, local authority work, housing association work, and so on. Whatever affects the funding of housing association work, doesn't affect industrial work. If you've got your feet in both camps, you're more insulated from changes in government policy or whatever. We find it difficult to do that and the majority of our work is in the housing sector anyway, we specialise in it. That's a problem for us; it's always in the back of your mind, the need to diversify your client base.

Market positioning is very much meshed with identifying and understanding particular clients, and this process was found to be proactive in nature. Consultant A, for example, makes extensive efforts

> to try and research, as much as possible, the clients from day one. The focus is on identifying and speaking to the right people within the client organisation, so that [the firm] knows exactly what the client really wants, rather than what the person from the client who [the firm] initially deals with wants.

Similarly, Consultant D asserts:

> [we] pride ourselves on our understanding of the needs of other design team members and have an informed view of the particular working practices of our clients, especially with regard to office space planning and catering.

This focus on better understanding particular clients' needs is consistent with the call for professional service firms to create greater 'client intimacy': 'delivering not what the market wants but what specific customers want' (Treacy and Wiersema, 1995: 35).

More specifically, the need for construction professionals to better understand the client and the client organisation is consistent with Walker (1989):

[the] professions . . . need to understand how their client's organisation operates
. . . the [professional] need[s] to have the ability to understand the structure of
their client's organisation . . . and in particular they should understand the
decision making mechanism of the client's organisation and where authority
for decisions lie[s].

(Walker, 1989: 12, 63)

Successful market positioning and client intimacy for small construction firms
was viewed as being particularly important, as each contract for a small
construction firm represents a greater proportion of its annual turnover compared
to a larger company. This is echoed in the literature, with Reid and Jacobsen (1988)
finding that many small to medium sized firms are dependent upon a smaller
number of customers for their turnover than large firms, and that these small firms
thus tend to be very close to their client base. The case study findings give clear
evidence that considerable effort is made to create and nurture business
relationships with repeat clients. Consultant B, for example, pointed out:

Generally speaking [we as the two directors of the firm] deal with the clients,
partly because we're absolutely dependent on them for work and therefore
can't afford to upset them.

This is consistent with the literature, which argues that successful small to medium
sized firms cultivate their customers closely, monitoring individual requirements
to keep their loyalty. This is done both at a social level and at an operational one
(Reid and Jacobsen, 1988). The different focuses of customer cultivation are
evident in the case study findings. Consultant A, for example, makes explicit efforts
at a social level, to

speak the same language as [our] clients; so, in a shared project with a client,
you are not talking about aesthetics; you are talking about funding; . . . [we]
even dress like our clients and don't wear tank tops . . . You have to be in tune;
you won't get a second chance if you haven't been able to respond in the right
way / speak the language, right from the beginning. The skill is actually being
able to decipher from these guys what is wanted and to be able to respond in
the right way.

While Consultant C develops its client relationships at an operational level by
having

contact at various levels with our clients. But on the ground, on site, our
surveyors, graduate or postgraduate surveyors, have contact with clerks of
works, people at that level. So they get a good relationship going, trust between
them, and that reflects on the firm . . . the way it works basically, and it goes

a lot with generations and age and anything else, is that my contacts with the clients are at a director level, if you like, senior. I don't know, now, the people on the ground or middle management. My surveyors do, because they're dealing with them everyday, and as they're getting older they're get promoted, clients are getting promoted. That's the way it works, it's a natural progression.

Consultant D identifies the potential problems of not developing appropriate client relationships at an operational level:

If the client has a very defined hierarchy, and everything must go through one particular person, and [one of the directors] speak to the wrong person they are in trouble. It's very difficult then to find a way through because, if you try and bypass someone, as [a director] has done in the past, to try and get information and the person bypassed finds out about it, then it can be counter productive because they get upset. So you have to be very careful about how you approach that.

In combination, the case study findings demonstrate that small construction firms understand their markets well, and, hence, how to position themselves. The imperative for small firms to identify and develop appropriate market positioning is consistent with the observation in the literature that:

small businesses cannot rely on the inertia of the marketplace for their survival. Nor can they succeed on brute force, throwing resources at problems. On the contrary, they have to see their competitive environment with particular clarity, and they have to stake out and protect a position they can defend. That is what strategy is all about – making choices about how you position your company in the competitive environment.

(Porter, 1991: 90)

The case study findings thus strongly suggest that both *business strategy* and *market positioning* provides appropriate focus for innovation activity, and is consistent with the distinction between 'strategic positioning' on the one hand and 'business investment' on the other (Mitchell and Hamilton, 1988).

Summary of business strategy and market positioning factors of innovation

In summary, the case study findings on business strategy and market positioning have three key implications for innovation in small construction firms. First, small construction firms are more exposed to the whims and movements of their business environments than large firms and, in necessary response, their business strategies tend to be more soft focus and reactive in nature. Second, the greater vulnerability to the market amplifies the need for careful positioning in multiple markets and client relationship development in order to spread the risk of variable workflows in any one market. Third, the dominant role of the owner(s) of small firms allows quick decision-making and innovation activity to take place in response to rapidly shifting market conditions and client demands; in effect, to create an agile firm. The very political strength that stimulates agility can, however, bring about an adversely myopic view of the 'best way' for the firm to operate.

Technology

The technology variable in the model is defined as the machines, tools and work routines used to transform material and information inputs (for example, labour, raw materials, components, capital) into outputs (for example, products and services). The case study findings suggest that technology takes three key forms: organisational work routines, information technology and knowledge management. This view of technology is compatible with the organisational definitions of technology. Hickson et al. (1969), for example, argue that technology is made up of operational, machine and knowledge technologies.

Work routines

Organisational work routines are stable patterns of behaviour that characterise organisational reactions to internal and external conditions. The case study findings indicate that routines within small construction firms exhibit considerable *stability at their administrative cores*, and significant *flexibility in managing information and resource flows at a project level*. This distinction is mirrored in the literature with Gann (1998), who separates business processes from project processes. Project processes are taken to be outside the traditional boundaries of the firm. They operate in a multi-actor environment, whereas business processes are considered to be more integrated with the firm itself. Gann stresses that the challenge is to analytically separate business and project processes in order to better understand them, but successfully integrate them in practice to produce seamless service delivery.

The relative stability of work routines in the administrative core is evidenced by such examples as Consultant B, who identifies that the routines to measure quantities and agree figures are fairly predictable:

> There's not a lot to learn. You're possibly learning about new forms of construction if you get a job that's slightly different than the normal bricks and mortar, perhaps a glass and steel building. Then you would be learning. If that type of thing came up again, then you'd have background knowledge, but the majority of jobs are pretty basic.

The routines reflect the repetitive nature of the tasks, with the significant computation aspects of this work being automated with 'four trainees who . . . just bash away on calculators all day and two or three people typing all day.'

Consultant C similarly suggests that although the work is organised informally, the day-to-day work has to go through a ISO 9002 driven set routine:

> If a job comes in, then it triggers a series of actions. At each stage the stages are recorded following a set pattern: things like confirming instructions, and filling in check boxes as each stage is completed, and not going on to the next task until that has been done.

Within this context of relatively stability, the principal focus of innovation in work routines within the administrative core is seen to be improving process-based efficiency and, in so doing, release organisational time and resources to enable progress of the business in other ways. Consultant A, for example, has set up a system of drawing techniques, using various standard components put together in different ways, which allows the firm to be more efficient.

The more volatile, multi-stakeholder context of the project environment is seen to be the driver for more flexible project routines to accommodate this increased complexity. The need for flexibility in managing project routines is demonstrated by the comment by Consultant B on project information and resources flows:

> We keep a record of how jobs are progressing, and then we're fairly flexible in how we respond to new information . . . we're aware of what's going on because we talk to people and say, 'Where are we? Have we got the information? Has somebody responded?'

Similarly, Contractor C explains, for example, that they do not think about operational innovation too much; usually a challenge occurs, they analyse it, and come up with ways of dealing with it – 'flexibility is paramount'. The challenge can be the result of client demands, workload, site conditions, or new technology and products that are put up in front of the firm. Contractor C stresses that construction projects tend to be unique in their requirements, and that:

> even if the specification for a fit-out contract is the same on different floors, the job won't be the same – you'll have to get the materials up to the higher floors. The programme often requires [us] to 'think and manage in a different way', so that [we] can complete works on or before deadline . . . [unlike site work] management and administration tends to follow established routes.

The finding that project work routines are flexible is consistent with the literature (see Chapter 2) which observes that construction innovation often takes the form of pragmatic problem-solving on site which could not be reasonably predicted before the project started (for example, see Construction Productivity Network, 1997; Winch, 1998).

Information technology

The case study findings suggest that the increasing investment in, and use of, information technology by small construction firms is an area of significant innovation activity in itself, and a powerful enabler for innovation in enhancing the quality and efficiency of the services they offer. The critical, and still emerging, role of information technology in improving both construction firm competitiveness and the quality of service provision is argued extensively in the construction literature (for example, see Betts, 1999).

Information technology within the small construction firms is considered to be focused very much on automating administrative and design functions and, to a more limited degree, to exploit knowledge better. Both focuses of information technology were viewed as being vital and rich sources of competitiveness. Contractor A sees itself as being increasingly dependent on information technology and views it as an advantage:

> and that is something seven or eight years ago I certainly would not have said. I fought against having computers for years, simply because I was not in the area and could not handle it.

Similarly, Consultant A identifies information technology as essential to sustained profitability, and stresses that it is striving to feed the use of technology into its 'whole philosophy and training, so that everything is technology orientated . . . there is [a] technological orientation to the firm.'

The automation and knowledge exploitation aspects of information technology is evidenced by the assertion by Consultant A that information technology speeds up the design process up and is viewed as being essential to remaining competitive:

> The beauty about AutoCAD, or any CAD system, is that once you have drawn something, you never have to draw it again. From one project to another, inevitability there are similarities and so you can copy a whole project across and alter/adapt it, as long as it is the same type of building.

The use of information technology to monitor work was also viewed as important in creating competitiveness. Contractor B, for example, has a proprietary software system with which it monitors projects:

> Each activity is assessed in terms of its percentage completions and then the data is input into the computer at the time the report is made. The system then throws out a graphical representation of whether the individual activities are behind or ahead of schedule. That then gives us an overall impression of whether the project is ahead or behind schedule . . .
>
> Not everybody does it at our scale of operation but clients see it and it makes them think they [the contractor] are in control, they do know what's going on.

The ability of innovative use of information technology as a unique source of competitive advantage which has the potential to differentiate its service from those offered by its competitors is illustrated by Consultant C:

> In the past, on a large-scale repair contract we would do the surveys by hand, translate the information into a document, which would be dictated and typed, then issued out as a text document. Nowadays, we do the surveys using hand-held computers, gather the same information in a more disciplined manner, import that into a database system, where it is validated and collected, reorganised, grouped. The numbers are crunched within the database, and the document is produced by that database. The database saves time in typing, and it improves accuracy in terms of transcription errors, and allows us to more quickly produce what-if scenarios for the client (what if we reduce this, what if we knock this house out, what if we add more heating in). It enables us to respond more quickly. That to us is an innovation that has happened in the last two to three years and is being constantly developed and refined. From my local experience of other consultants working for the same clients as ourselves we are the only consultants using a system anything like that.

The focuses of information technology as both an area of innovation in itself, and as an enabler for innovation resonates strongly with Barrett and Sexton (1998), who view information technology as having three potential roles: administrative, operational and competitiveness generating. The administrative role appreciates the powerful capability of information technology to automate, for example, accounting and control functions. This role requires the design and implementation of an effective and efficient information technology platform and is relatively independent of the strategic management of the firm. The operations role is an extension of the administrative role and is distinguished by the creation and management of an information technology platform that allows the automation of all business processes, not just administrative processes. The competitiveness-generating role, in contrast to the other two roles, is more strategic in orientation, and is a role which is generally under-utilised by firms. Rather than harness information technology for internal efficiency, a more strategic perspective views information technology as an innovative means of leveraging organisational knowledge, information and technological competencies to obtain different sources of competitive advantage.

Knowledge management

The case study findings demonstrate considerable innovation activity in small construction firms in developing their competency in sharing and exploiting knowledge. This is achieved either by using information technology (see 'Information technology' section pp. 49–50) or through people-to-people or more paper-based knowledge repositories and systems. It is seen as a rich source for improving service quality, and as a means of better utilising human resources. Knowledge management can be defined as

> the discipline of creating a thriving work and learning environment that fosters the continuous creation, aggregation, use and re-use of both organizational and personal knowledge in the pursuit of new business value.
>
> (Cross, 1998: 11)

Consultant A views information management and sharing as important, and makes extensive use of databases, proposal documents, meetings, review strategies, etc. The focus of this firm's action research innovation (see Chapter 3) was to develop and implement a knowledge repository to improve the efficiency of a particular design activity.

One of the architects in Consultant A has a particular specialism in industrial design and procurement. The innovation concentrated on how to capture and codify this tacit knowledge in such a way that it would become a shared capability which other architects could use. Tacit knowledge is defined as

> something not easily visible and expressible. Tacit knowledge is personal, context-specific and hard to formulize and communicate . . . Subjective insights, intuitions and hunches fall into this category . . . [it] includes cognitive and technical elements.
>
> (Nonaka and Takeuchi, 1995: 56, 73)

The aim of this capture and sharing of tacit industrial design and procurement knowledge was to satisfy the 'immediate need to delegate and have confidence that others can carry out the given task'. The architect noted that:

> now there is more and more competition [and there is a] need to limit the length of the learning curve; fill in steps and get them there quicker. It can be duplicated, so why not exploit it? With the system, people will be able to do the work in a quarter to an eighth of the time.

Various pockets of documentation and knowledge were brought together into a single manual which focused on the inception to feasibility stage (a third of the overall design process). The system

> set out how the knowledge goes together as a structure: the materials involved, the procurement process, down to the nuts and bolts. [In particular] a lot of it [is] strategic planning: the floor space, contract procurement, grid sizes and the services needed to be provided, for example. These are the global perimeters. It is to do with design.

The sharing of knowledge was deemed to allow the architect to

> get on with other things. Initial evaluation of the system proved positive. A new architect, who had no previous experience of industrial design, used the system for four or five different layouts on three sites, completing the work in approximately 40 per cent less time than normal.

This capturing of tacit knowledge into explicit knowledge is consistent with a current trend in knowledge management to migrate knowledge from human capital to company capital. Human capital is made up of the skills and experience of employees (Sveiby and Lloyd, 1987). Company capital includes such elements within a firm as its administrative systems and its culture (Sveiby, 1997). (Note: for this report, the term 'company' capital has replaced the original label of 'internal structural' capital.) The migration from human capital to company capital enables knowledge to be embedded within the technology of the firm, facilitating knowledge sharing, and mitigating against the risk of losing knowledge when staff leave.

The intended aim of such knowledge storage and sharing to release senior professionals' time is also in accord with the concept of leveraging human capital, or the delegation of semi-routine and routine tasks by senior to junior professionals (Sherer, 1995). This leveraging allows limited senior professional resources to be deployed in superior profit generating activities, enables the firm to expand its service capacity without diluting firm profits, and some of the efficiency gained by leveraging human capital can be passed on to clients.

Summary of 'technology' factor of innovation

In summary, the case study findings on the 'technology' factor of innovation have three key implications for innovation in small construction firms. First, information technology is an increasingly important *focus* for innovation in itself, and as an *enabler* for innovation. Second, 'soft' technologies are particularly important for small construction firms. The required 'soft focus' and 'agility' of small firms to compete necessitates work routines which provide stability at the administrative core, and flexibility in project processes to adapt to rapidly changing market conditions and client needs. Third, knowledge management is seen as a way to transfer tacit knowledge located in individuals to company knowledge. This is particularly important for small construction firms, as often a significant proportion of their knowledge about clients and work activities is embodied in a small number of individuals. This renders small firms especially vulnerable to key members of staff leaving the firm.

People

People are viewed as possessing knowledge, skills and motivation to perform a variety of tasks required to do the work of the firm. This element of the organisational innovation model echoes the competence literature (for example, see

Cohen and Levinthal, 1990; Hamel and Prahalad, 1994). Nordhaug (1993: 50), for example, adopted an explicit human capital perspective of competence, and defined it as 'the composition of human knowledge, skills and aptitude that may serve productive purposes in organizations'.

The case study findings identify the ability and motivation of staff to be a critical resource for innovation. More specifically, ability is considered to be made up of aptitude, training and experience; and motivation to be a function of both extrinsic and intrinsic rewards. (Extrinsic motivation is brought about by external rewards, such as pay and job security, while intrinsic motivation is from internally generated rewards, such as enjoyment of tasks, achievement and recognition: Herzberg, et al., 1959.) These general conclusions are consistent with Lawler (1973), who argued that performance was a function of ability and motivation, and that ability was, in turn, a function of aptitude, training and experience. The project findings extended the idea of people as a valuable resource, by identifying that the flexible management of people is a key dynamic capability for innovation.

Strategic importance of people

The strategic importance of staff for generating competitiveness is evidenced in the following observations:

> The people in my company are very important, particularly the good ones, because we are so small.
>
> (Contractor A)

> the quality of staff is one of the firm's most important assets . . . we can be as clever as you like, we can market, we can sell, we can be as dynamic as you like, but if you don't get on site and produce a quality finished product, then, one, you don't get paid, and, two, you don't build a reputation.
>
> (Contractor C)

The pivotal role of having the appropriate quality of staff to generate and implement new ideas is identified by Consultant A, who explains:

> This is absolutely key . . . the quality and training of staff, not just key members, but key members even more so . . . it comes down to the ability of the people and the motivation.

Ability of staff

A common problem identified by the small construction firms is the difficulty in attracting staff with the appropriate aptitude and ability to undertake work and to drive innovation. Contractor A states that there is a dearth of capable and motivated supervisors and labourers coming through into the labour market: 'most of the young ones don't know how to work and don't want to work'.

Similarly, Consultant B notes having a problem with finding suitable trainees:

> They tend to be 18, with A-levels, partly because we can't find people at 16. We would prefer to take people on at 16 and bring them through college – ONC, HNC, degree course. We tried last summer, but were unsuccessful in finding anybody with a spark of life or interest. We advertised, we tried the TECs, the colleges, schools, without success.

The strategic importance of attracting appropriate staff to open up new markets is also noted, with Consultant A commenting that: 'we are currently looking for people with the right expertise to enable us to enter the residential market'.

The need to develop the ability of staff through training is supported by the small firms. Contractor B comments, for example, that:

> Most organisations have agents who are trade-based, and have worked through the company, have competence and experience, and have proved themselves. Our lads get trained as I've been trained so that they can do almost everything: programme, resource. Know expected outputs from men and machines, know costs of men and machines.

There was consensus, however, that the most appropriate form of training was 'learning-by-doing', rather than by external training. The reasons are twofold. First, external training was generally considered too expensive. Consultant B explains that although the chartered surveyors within the firm attend continued professional development events to satisfy the (Royal Institute of Chartered Surveyors (RICS)), they are less willing to commit resources to other training:

> I think if [staff] were prepared to take a day's holiday and pay for it themselves there would be no problem at all. I think if someone said I'd like to go on a day course on the new form of contract the answer would probably be no. That's because of the cost of the course and the time out of the office. The courses tend to be in the region of £200 per day plus the lost time as well.

Second, there is the difficulty in transferring what is learned from external training into the workplace. Consultant C provides insight into this issue by saying:

> The cost of training is a barrier. The cost of training someone on Access might be £500 for a two or three day course. The value is that they will come back knowing what they can do. They will then want to actually implement that and put it into practice, but because they've got other things to do, they won't have the time, and as a result they'll forget what they've learnt. The only way to be competent and efficient is to work at it a lot. So the barrier to training other people to do it is the fact that we don't need to, because we've got people who know how to do it.

The importance of learning and development in the workplace leads into the third reason that external training is not given a priority, namely, that development through experience is considered to be the best form of training. Contractor A, for example, argues that although off-site training is resourced by the Construction Industry Training Board (CITB) to provide training for its bricklayers and joiners, it is felt within the firm that on-site training is more valuable. Similarly, Consultant C articulates that:

> Experience is more important in our environment than training, in the sense of someone on a one-day course or three-week course, which is nowhere near as valuable as three months' experience of doing a job. The rare commodity, the most expensive commodity, is experience . . . What we need are people with a lot of experience. That experience rubs off onto people that have less experience, the more so because everyone works physically quite close together, so no one's working in isolation and they've always got people to help. Having people, associates and senior level of staff, who are very experienced and competent at their job raises the level of everyone else. Think the way we work together assists this . . . so when a graduate first goes into that environment he tags along basically with a senior guy. If the senior guy sees that he's doing all right he'll say, 'OK you go and do that, and come back.' You know, it's just a gradual process. As we get more and more confident in them and they get more and more confident in themselves, then they get more jobs to do; they're given more responsibility and that's how they grow. In the end they do it themselves.

These findings are consistent with Banfield et al. (1996), who concluded that the owners of small firms often did not recognise the need for training because of scepticism concerning the value of training for small businesses and an unwillingness to commit the necessary time and resources.

Motivation

The case study findings highlight the importance of motivated staff. Contractor B, for example, comments:

> We need self-motivated people, people who are happy to get on with the job and know that they are being appreciated.

Two principal issues concerning the appropriate motivation of staff to create and support innovation are identified in the findings. First, very much aligned to extrinsic motivation, there is the role of career structures and progression. Second, there is the importance of an appropriate organisational culture in motivating staff in an intrinsic fashion.

Taking the first issue, the small firms are characterised by fairly informal career structures and pay scales. Consultant B, for example, notes:

> Obviously there is a financial incentive to progress. But it is not a structured progression if you like. There aren't any Quantity Surveyor Level 1, Level 2 etc., it's very informal, because we're a small firm.

The informal nature of the career structure and the inability to pay the same type of salaries to staff as large firms is viewed as sometimes having a detrimental effect on retaining high calibre staff. Consultant C explains:

> We recruit graduates on a regular basis. If you get a very bright one and you want to keep him then you've got to give him an incentive to stay. But if he's very bright and wants to make a lot of money then he'll see straight away that his prospects are very limited here, because of the size of the firm and because we're a three-partner partnership. The only way he can grow is to become a partner and that's not always possible. So they tend to move on. Some leave because of the type of work we do; we can't always give the broad range of experience that these young lads want and need to satisfy their diary requirements for the professional qualification. We only have a number of opportunities to do that and provide them with that specific experience, we have to select jobs for them.

The difficulty of recruiting and retaining key staff in construction profession service firms is supported in the literature. Barrett and Ostergren (1991) map out the potential implications of losing key staff, and emphasise the importance for retention of designing appropriate strategies to create and nurture high levels of motivation in staff.

The case study findings indicate that the small construction firms address the inherent informality of the extrinsic side of the motivation equation through the development of a nurturing culture that fosters a sense of community and belonging. This position is portrayed by Consultant C in the argument that:

> The culture of the firm flows from the client relationship. I suppose you have to have a professional culture. In a firm like this we try to generate a culture of hard work. I think it's slightly different in a small firm. You can't get away with playing computer games during the day and things like that. We can't afford to have passengers in this firm. The culture of the firm is the work ethic and the quality side of things. But we also try to make it a friendly working environment, because it's a relatively small firm and that is important to a lot of people. If they like the working environment they stay. They see themselves as part of a firm; it's more of a family business than a company.

Similarly, Contractor B explains that the firm's culture is

> not a fear culture; it's a positive culture I'm trying to engender [in] everyone . . . I hope it'll prove beneficial to create a positive environment that we can

all enjoy to work in. I mean, we all need to pay the bills, but you can pay the bills with a smile on your face, or you can pay the bills thinking 'Thank God it's the end of the day'.

The need for an appropriate culture which stimulates and supports innovation is documented in the literature (for example, see Peters and Waterman, 1982; Kanter, 1984). Tidd et al. (1997) argue that a creative culture

> involves systematic development of appropriate organizational structures, communication policies and procedures, reward and recognition systems, training policy, accounting and measurement systems and deployment of strategy.
>
> (Tidd et al, 1997: 326)

Flexibility of staff

The case study findings demonstrate that the firms develop multi-skilled staff and manage their staff in a very flexible fashion to ensure that they can successively focus their activity to meet varying market demand and special requirements from clients. This flexible, multi-skilled dimension is illustrated by the following observations:

> There's a realisation these days that if there's a job to be done it has to be done. They've even had me [a director] measuring when they're busy . . . I think that with any small firm people have to turn their hand to whatever is the panic of the day. Even the administrators, to a limited extent; both of them are capable of using spreadsheets and things like that so occasionally can be brought into the quantity surveying process, the actual technical process is very simple.
>
> (Consultant B)

Similarly, Consultant D does not assign certain areas of work to staff, they

> purposely mix it up all the time so, for example, one month someone may be designing a hospital project, the next month an office or school. There is a continual flow of information like that, so if you say you are used to doing fast turn around projects within an office environment, you are going to take those same ideas and use them in say a hospital environment i.e. carry the ideas forward. If you work in the same area all of the time . . . after a time you have no idea about the new things happening in other areas, because you have never come across it.

The flexible nature of task allocation is supported by managers having an in-depth knowledge of all of the staff's abilities. This is facilitated by the small number of staff. Consultant B, for example, commented:

I think because most of the people have been here a long time you get to know their strengths and weaknesses and therefore that teaches you how to deal with any particular problem or thing that happens. You know that certain people, as we all do, have strengths and weaknesses. One of the associates has done a lot of work on the prison service, so if it's a prison job we tend to use him, because he knows the forms and the way they work, the people. If he were desperately busy we would leave it to one of the others, who are capable but a little bit slower and would need some assistance perhaps.

Summary of 'people' factor of innovation

The case study findings on the 'people' factor of innovation offer a number of important implications for innovation in small construction firms. First, the appropriate ability and motivation of staff is paramount for firms to create, manage and exploit innovation. Second, staff need to have a broad range of skills and experience to undertake multiple tasks. This flexibility is especially pertinent to small firms, who need to be 'agile' with limited, and often very stretched, staff resources.

Organisation of work

The organisation of work variable in the model is defined as the creation and coordination of project teams and commercial networks both within the firm and across its business partners. The collaborating firms are project-based firms, in that their work consists primarily of projects (for example, see Artto et al., 1998).

The case study findings indicate two important dimensions to the organisation of work: the allocation of work down the hierarchy from the client into the organisation, and the resourcing or doing of project work.

Allocation of work down the hierarchy

The case study findings evidence a fairly consistent pattern of how small construction firms secure work, and allocate and coordinate this work within the firm. First, the owners of the firm secure work normally through direct contact with the client. Second, the work is allocated to a project manager who organises and manages the project to its completion. Due to the size of the firm, the owners and the project manager can be, on occasion, the same person. This pattern was evidenced by the construction firms in the following observations:

one of the directors would allocate a project to usually one of the associates, unless it's something very specific, a specific client perhaps or a specific type of work, in which case [one of the directors] might keep it to [themselves].

(Consultant B)

it's just like any organisation: there are layers of management ... at the moment, for example, I've got a team of surveyors up in Aberdeen doing some survey work and basically we [the partners of the firm] set out the parameters, the targets, we do an analysis and then we give them the tasks. One is to appoint a senior team leader and he would organise that at a local level in terms of travelling and who does what, targeting, checking etc.

(Consultant C)

This is consistent with Maister (1993), who suggests that the division of labour in professional service firms comprises finders, minders and grinders. Finders are responsible for finding new clients in addition to nurturing existing client relations. Minders coordinate organisational and project work. Grinders carry out the majority of the routine work.

Project teams

The reporting and communication structures of project teams are very much set up as part of the allocation of work down the hierarchy (see subsection above), with clear project management responsibilities for planning and coordination of project work being allocated. Contractor C, for example, explains:

If there is an enquiry, then the managing director or another director will try to follow that through to tender submission. Actual jobs tend to be allocated on a project-by-project basis, according to the availability of staff, and other factors, such as the size of the contract, pre-planning, lead-times, and the job. A director will liaise with the client about financial issues; and an office-based contracts manager will oversee the on-site construction work.

The resourcing of projects and the organisation of project work is flexible, depending on workload pressures and availability of staff and skills. Consultant C, for example, comments:

I think there's a lot of people with diverse skills in the firm and they do mix and match and change around all the time. I've got guys out doing surveys who could be in the office, at a PC, working on CAD; or they can go out, do some measured surveys, come back and plot them on CAD. I've had guys who're doing some maintenance inspection work. They went out, did the surveys, came back and prepared a spreadsheet and a database, which listed all the items of work, costed it, and made the comparisons, and produced a final report. So they are very flexible.

Summary of 'organisation of work' factor of innovation

The case study findings on the 'organisation of work' factor of innovation identifies a number of key implications for innovation in small construction firms. There is a fairly uniform pattern of allocation of work down the hierarchy process that establishes reporting and communication structures at corporate and project levels. The resourcing of projects and the organisation of project work is flexible in response to limited staff resources and the volatile and unpredictable characteristics of construction projects.

Summary

The organisational model of innovation has been introduced and discussed. The principal focus and utility of the model is that the generation and implementation of a new idea to enhance overall performance requires the practitioner to take a broad, systemic view. Innovation in any one area of the model needs to be understood and organised in conjunction with the other factors. The introduction of new information technology, for example, may very well have significant implications for 'people' (such as, having the skills to fully exploit the technology), 'organisation of work' (in terms of, say, new business and project processes), and 'business strategy' and 'market positioning' (for instance, the potential to offer new or improved services).

The following chapter illustrates how this innovation model can be used in practice.

Chapter 6

Context of innovation

Introduction

This chapter explores the context of innovation. In particular, it seeks to address the following questions set out in Chapter 2:

- What are the key events external and internal to small construction firms which trigger innovation activity?
- What is the appropriate emphasis between market-based innovation and resource-based innovation in small construction firms, and what conditions dictate this emphasis?

What are the key events external and internal to small construction firms which trigger innovation activity?

The case study findings identify two principal modes of innovation (see Figure 6.1) which is useful in both understanding and managing the interaction between market conditions and the firm. The two *modes of innovation* are shown in the centre portion of Figure 6.1. Mode 1 innovation focuses on progressing single project, cost-orientated relationships between the client and the firm, and Mode 2 innovation concentrates on progressing multiple project, value-orientated relationships between client and the firm.

The right-hand side of Figure 6.1 reinforces the notion that the mode of innovation is substantially determined by the nature of the *interaction environment*: an enabling interaction environment encourages Mode 2 innovation, and a constraining environment is conducive to Mode 1 innovation. An enabling interaction environment is one which the firm can influence to a significant extent or is relatively stable, *enabling* the firm to innovate within a longer term and more secure context. A constraining interaction environment is one which a small construction firm can influence only to a limited extent or is relatively unstable, *constraining* the firm to innovation activity undertaken within a shorter and more insecure context.

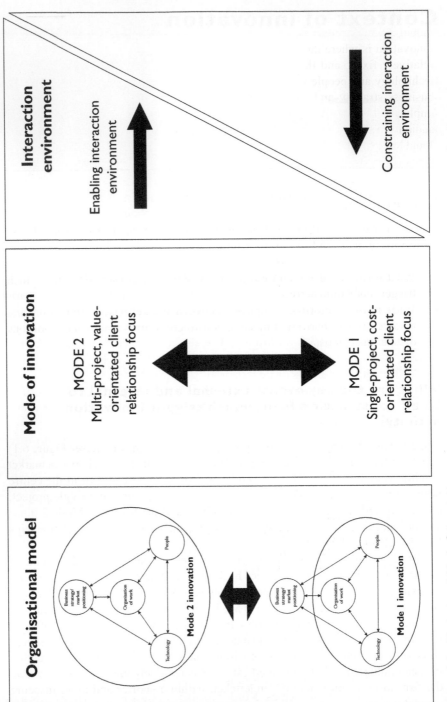

Figure 6.1   Mode 1 and Mode 2 innovation

The left-hand side of Figure 6.1 identifies which factors of the organisational model are the primary focuses of (and levers for) innovation activity: Mode 1 innovation is where the business strategy and market positioning variables are relatively fixed, and the focus of the activity is in the organisation of work, technology and people variables; Mode 2 innovation involves innovation in the business strategy and market positioning variables which, in turn, will have implication for the remaining variables. It is evident that Mode 1 and Mode 2 are addictive in nature, i.e. Mode 2 innovation encompasses Mode 1 innovation. Mode 1 and Mode 2 innovation presents a continuum, rather than a choice of two discrete types of innovation activity, i.e. 'hybrid' modes of innovation can be located between Mode 1 and Mode 2.

The 'modes of innovation' model resonates with the single and double loop learning literature. Single loop learning is where an organisation adjusts its behaviour relative to fixed goals, while double loop learning is where an organisation and its goals are open to change (Argyris and Schön, 1978). Mode 1, or single loop innovation, is where the business strategy and market positioning are fixed, and the firm adapts the organisation of work, technology and people within this context. Mode 2, or double loop innovation, is where the business strategy and market positioning variables, i.e. the context, are open to change.

What is the appropriate emphasis between market-based innovation and resource-based innovation in small construction firms, and what conditions dictate this emphasis?

In Mode 1 innovation, small construction firms view markets as being highly competitive, with little scope for negotiation or collaboration between firms and clients with respect to the client's brief. The clients are concerned primarily with reducing costs and/or delivery times, and tend to favour price-competitive procurement methods. This position is evidenced in the following observations:

> persuading the client that they should be interested in quality and longevity is difficult . . . [and that] on traditional design and build all the contractors are really interested in is building at the cheapest price, so that innovation suffers.
>
> (Consultant A)

> cost is a problem because clients want to spend as little as possible, and don't consider lower energy costs or lower maintenance costs. This is often because there are restrictions that they are under, in terms of budgets, and the distinction between capital works budgets and maintenance budgets; they can't afford to pay now.
>
> (Consultant D)

contracts are won through compulsory competitive tendering; lowest price wins.

(Contractor A)

compulsory competitive tendering puts pressure on you to put out a low bid.

(Contractor C)

This price focus is also combined with what is perceived by the small construction firms to be a generally conservative approach by clients to innovation. Consultant B, for example, remarks:

public clients, such as the health services, won't let you be innovative, unless there is a demand for it; they want to go down set routes, because of the requirement of accountability. Clients don't like too much risk.

Similarly, Contractor A comments:

clients have conservative tastes; they tend to want things that have been proven; the clients are working to budget and the budgets are generally three years behind, so they can't afford to change. If anything designs are more traditional now.

For their part, the firms tend to be unable or unwilling to set themselves apart from other firms in a critical way through innovation. Firms in Mode 1 are therefore both *constrained* by lack of market encouragement and/or internal motivation and capability to innovate. This is illustrated with the following observations:

we don't sit down and decide we are going to be very different because of this and that. There are too many established procedures for us to buck that, for example, the tender process, the way that work is awarded, the way that client choose clients. We have to go with the flow.

(Contractor C)

It is and it isn't [reputation]. You just have to be in the right place at the right time or know the right person. Because the work that we do, to be quite honest, it doesn't really matter, whether you are good, bad or slow. It is local authority and housing association and they have a list and they have to stick to that list. It doesn't matter if you are £100 dearer than the next. The next could be the worst contractor on earth but people will want them. Reputation doesn't come into it.

(Contractor A)

Innovation by firms in Mode 1 is focused primarily on maintaining existing clients in known markets through project-specific 'problem-solving' innovation and incremental improvements in the organisation of work, technology and people variables of the organisational model of innovation. Consultant C, for example, comments:

> If we have an innovative system that allows us to be more efficient, then we can pass those savings on to the client in lower tenders and actually win the work that we might not do in other circumstances. For example, I did a presentation to a potential client in Glasgow who wanted some stock survey work doing, and I demonstrated to them a system that we have developed, and that was innovative when we produced it. I was one of about six consultants who made a presentation – the others were from big national companies. When the tender document came out they asked us to name two demonstration sites that we had in Scotland. We had one, but it wasn't up and running, and was a very small job. So I wrote back and declined to tender. They then contacted me and asked me to tender because they thought that the system that we had demonstrated to them, particularly the statistical analysis modules, was the most innovative that they'd seen, and the selling point of the system was that because of the way we do the statistical analysis they could get better results from a lower sample and therefore save themselves money. We are tendering for the job.

This issue is developed by Consultant B in the observation that:

> clients are looking more quickly, to get on site at an earlier date. To do everything that much quicker and cheaper; demand is geared to on-site capital spend and the speedy occupation of buildings. The development of techniques has been basically speed driven, rather than quality driven.

Similarly, Consultant A emphasised that the site-specific nature of projects stimulated innovation. The case of using prefabrication for the steel structure and cladding on a restricted city centre site was cited as an example. Mode 1 innovation is illustrated in Box 6.1. In summary, Mode 1 innovate stimulates incremental project-specific innovation geared towards generating cost efficiency while maintaining quality. The fragmented nature of Mode 1 innovation makes it hard to capture, integrate and progressively leverage the firm as a whole. This aspiration is the focus of Mode 2 innovation, and is discussed below.

In Mode 2 innovation, firms have arrangements with clients whereby the client offers continuity and predictability of workload for a significant period of time which, in return, provides a more secure, longer term context which *enables* firms to innovate and to provide the client with a competitively superior service. Mode 2 innovation is focused primarily on improving business arrangements through innovation in the business strategy and market positioning variables of the organisational model of innovation, which often has significant, systemic implications for the remaining variables of organisation of work, technology and people. Mode 2 innovation is illustrated in Box 6.2.

Box 6.1 Contractor C partition innovation

Business strategy

Contractor C is principally a fit-out contractor. The long-term vision of Contractor C is to build the turnover to £100 million, divided into thirds between property development, negotiated design and build, and traditional tender work.

Market positioning

Contractor C competes in a competitive fit-out contracting market. One of the projects the firm was involved with was the refurbishment of the headquarters of a bank. The job had very challenging demands: completion in the first quarter of the financial year, normal working in the headquarters to be maintained, requiring work to be done outside of office hours, and the highest quality finishes. The principal problem Contractor C faced was that the glass panels specified by the architect would take six weeks to manufacture, and the manufacturing process could begin only after the framework had been installed. Normally, Contractor C would demolish the existing framework, install the new framework, and use the lead-time to put carpets in and decorate.

Technology

The innovative solution to the lead-time problem was:

> to build the thing backwards ... we've decided to install the new framework first, and then demolish the old framework the day before the new stuff arrives.

Organisation of work

The partition innovation involved working closely with the design consultant and the client team, and interpreting novel design information:

> [the design consultant] produced some very sophisticated three dimensional visuals, computer-generated stuff. The client expects this level of service and has got its own team of professionals on board who have got a watching brief. They're scrutinising all the designs and costs. It's a first for us; we've never seen such sophisticated design information

produced for the client. I understand this is pretty much leading edge. For the size of contract we do, you don't normally get stuff like this produced . . . my role is one of interpretation; of taking that, going down to site, actually transforming this.

People

To ensure the innovation was successful, Contractor C saw that as well as working closely with the designers and the client's professional team (see 'Organisation of work' above) strong project supervision and senior management contact with the client was needed throughout:

> We will be live on the site next week. I've got a site supervisor lined up, who I will work with. Now I'm going through with the project for various reasons. It's quite a prestigious client; it's a difficult site; and the expectations of the client are high. I feel that we have got to supervise that from a fairly senior level. I want to maintain this client, and I see a lot of work coming forward . . . I'll run with it closely, closer than most projects. I'll literally be there every two or three days, because there's some tricky one-off designs that I need to relate to on site. I'm anticipating maybe a few hiccups, so I don't want to pass the parcel to somebody.

Outcome

The successful design and implementation of the partition innovation substantially contributed to the project being delivered within the tight time and working constraints.

In October 2000, Contractor C was awarded the first prize in the £100,000–£250,000 contract-value category by the Association of Interior Specialists for its fit-out of the Bradford and Bingley Headquarters. The principal of Contractor C believes that the award was given in recognition of the innovate approach to the installation of the partitions.

Box 6.2 Consultant C partnering innovation

Business strategy

In 2000, Consultant C successfully negotiated a three-year partnering arrangement, worth £6 million to £12 million per year, with one of its largest clients, a regional housing association. The arrangement started in April 2001. It involves the provision of a range of building surveying services, at agreed prices, to all six geographical divisions of the association, in return for which the client will make regular monthly payments, reconciled periodically with actual work. The services will include such things as disposal surveys, dilapidation surveys, audits and option appraisals.

Normally the firm's appointments with clients are one-off, so that it is 'living from hand-to-mouth all the time'. The new arrangement is perceived by the firm to have many advantages:

- it will guarantee over 50 per cent of the firm's annual fees
- it will provide work for the survey department and the contracts department
- it will enable the firm to plan in terms of recruitment and in terms of equipment, with more confidence than before.

The senior director commented:

> The innovative thing, perhaps, was thinking that we could do it in the first place, which was something that we had not done before, a national contract. I think the partnering arrangement is very innovative. It's the first time that we have ever had an arrangement like this with the client.

Market positioning

The partnering arrangement has advantages for the client over its previous arrangements for survey work. Consultant C will do a rolling survey of the client's stock, which will help the client to plan the maintenance and renewal programmes, including the replacement of worn-out elements such as kitchens, bathrooms, and heating systems, more efficiently and effectively. This will include the use of specialist option-appraisal techniques. It will also save the Housing Association about £750,000 over three years, and improve its cash flow. On the latter point, Consultant C explained that the association no longer funds its work through government grants:

> What they have to do now is mortgage the stock, raise money on money markets. They go to the merchant banks and say, 'I've got 8k properties, generating X million pounds in rent, I want to borrow X million quid'.

When that happens the client has to provide a lot of information to the merchant banks. Consultant C will help the association to do it because the firm will be doing a rolling survey of the client's stock: 'This information that we generate gets packaged and is part of that bid process.'

Technology

A central focus to successfully implement the partnering innovation is the development of the firm's information technology systems:

> We've got lots of little packages that we've developed on an ad-hoc basis in the past for little jobs; but this is a three-year job, and we'll be doing lots of development work on the IT side. We should have two people working full-time on that, and that motivates them because they're working on something new . . . we will have our feet so firmly under the table that no one else will be able to step into or shoes, if we provide them with a quality service and they get used to it and like it . . . [besides] there isn't another consultant in Merseyside who can do this anyway. They can do the building side bit, but what they can't do, and what we have specialist expertise in, is the database modelling.

Organisation of work

To meet the challenge, the firm has realised that it will have to change its management methods and reorganise the way they run things. The firm has decided to restructure the firm into nominated teams, with central points of contact, to deal with each division of the client. They are looking at the possibility of direct information technology links with the client, other than email: 'We'll need an ISDN line or something so that we can dump data backwards and forwards more quickly'.

People

The directors think that:

> [the partnering innovation] will involve some additional staff training, specifically database training . . . we're going to have to reorganise this, we're going to have to train, we're going to have to give people experience, knowledge, we're going to have to alter our structures and processes, to a certain extent, to match the divisional structure of the client, and the response time, because the client doesn't want to speak to a partner that hasn't got direct hands-on experience of the job, he'll want to speak to the team leader, who's dealing with Leicester or whatever. They're going to set up [an internal Consultant C] user-group, and there's going to be feedback, and I'll be going to that and so on.

The 'modes of innovation' model advocates a synthesis of complementary schools of thought on how firms shape or are shaped by their environments. The positioning dimension to the model is very much located within the industrial organisation literature which focuses on the selection of industries and the positioning of firms within that industry to achieve sustained competitive advantage (for example, see Porter, 1980, 1985). Further, the managerial challenge is contingency theory based, in that the task is to achieve appropriate fit with the environment. This is consistent with the contingency-based view which advocates that the degree of change or uncertainty in a firm's environment determines the optimal organisational structure (for example, see Burns and Stalker, 1961; Lawerence and Lorsch, 1967; Donaldson, 1988).

The industrial organisation and contingency perspectives emphasise market context over firm strategy, thus discounting or ignoring the opportunity firms have to shape their environment. The 'modes of innovation' model overcomes this limitation by being explicitly linked to the organisational model of innovation which integrates the managerial choice or strategic choice perspective (for example, see Thompson, 1967; Miles and Snow, 1994; Child, 1997). This perspective argues that firms are not always passive recipients of environmental influence but also have the opportunity and power to reshape the environment. The implication of strategic choice theories for firm innovation strategy is that management should take into account the multiple ways in which firms interact with their environments through the process of mutual adaptation between the firm and its environmental domain. The synthesis of industrial organisation, contingency and managerial choice perspectives have led to a more inclusive view of contingency theory which builds upon the assumption that 'increased effectiveness is attributed to the internal consistency, or fit, among the patterns of relevant contextual, structural, and strategic factors' (Doty et al., 1993: 1196), that is companies which have a fit between and within these factors will perform better than companies which have one or more misfits. Research into construction professional service firms has similarly demonstrated that firm profitability is contingent upon an appropriate balance between technology, people, structure and the environment (Barrett, 1995).

Summary

The case study findings identify Mode 1 and Mode 2 innovation, which are substantially a function of whether the interaction environment is *enabling* or *constraining*. Mode 1 innovation is principally concerned with innovation in the 'organisation of work', 'technology' and 'people' variables of the organisational model of innovation, with the 'business strategy' and 'market positioning' variables being relatively fixed. Mode 2 innovation focused on innovation in the 'business strategy' and 'market positioning' variables of the organisational model of innovation, with associated implications for the remaining variables of 'organisation of work', 'technology' and 'people.'

The key implication for small construction firms is that they should not 'flip' from Mode 1 innovation to Mode 2 innovation. Small construction firms need to incrementally nurture, or identify and move into, supportive enabling interaction environments. This is achieved through careful and integrated consideration and development of all the variables in the organisation model of innovation. It would be potentially disastrous, for example, for a small firm to enter a partnering relationship with a client without the necessary 'organisation of work', 'technology' and 'people' to fully satisfy client and firm needs and expectations.

Whatever the mode of innovation, innovation does not just happen in a small construction firm; rather there is a process of innovation. This issue is discussed in the next chapter.

Chapter 7

Process of innovation

Introduction

This chapter seeks to identify the process of innovation in small construction firms, in particular, by using case studies to address whether the process is rational and/or behavioural in nature.

Are the processes of innovation in small construction firms rational and/or behavioural in nature?

The case study findings reveal that the innovation process follows an iterative, nonlinear cycle of phases, as shown in Figure 7.1.

The process has five parts:

1   *Diagnosis*, where the issue (be it an opportunity or problem) forming the focus of the innovation activity is identified, and information is collected for a more detailed diagnosis.
2   *Action plan* is prepared after the diagnosis, where possible ways to progress the innovation are developed, and from which an agreed plan of action emerges. This provides the basis for the next step.
3   *Taking action* is when the idea is put into practice.
4   *Evaluation* next takes place to determine whether the innovation has been a success or not.
5   *Specific learning* is then undertaken, where innovation is reassessed, areas for improvement identified, and the process begins another cycle.

Innovation activity does not take place in five sequential stages; rather, as depicted in the outer ring of Figure 7.1, the cycle can take place at each stage of the 'overall' process. At the diagnosis stage, for example, the practitioner might well go through periods of evaluation and reflection to confirm that the 'innovation gap' identified is appropriate. This 'overall' process continues until the innovation is either successful, or it is decided that the innovation is not appropriate, and should not be continued.

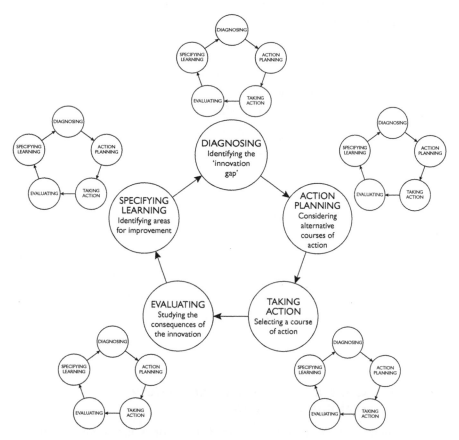

Figure 7.1   The process of innovation

The process of innovation is not uniform; rather, as shown in Figure 7.2, the process is characterised by an interplay between forces of 'action' and 'reaction' over time which progresses or hinders the closing of the 'innovation gap' between the current level of performance and a desired level of performance. There are myriad potential action and reaction forces in each of the organisational innovation variables, such as, strong senior management support for the innovation (action), resistance to change from staff (reaction), allocation of capital to purchase needed technology (action), and lack of appropriate work routines to coordinate and channel the innovation activity (reaction). Figure 7.2 presents a situation where the innovation is successful, i.e. the 'action' forces, over time, have overcome the 'reaction' forces. It is just as feasible, of course, for the 'reaction' forces to be stronger than the 'action' forces, and for the innovation to fail.

Box 7.1 provides a mini case study which illustrates the various dimensions of the process of innovation discussed above.

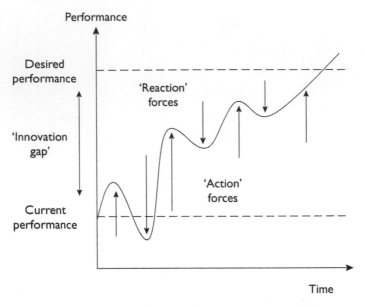

Figure 7.2 Action and reaction interaction in innovation

Box 7.1 Contractor A health and safety training innovation

The action research innovation which Contractor A undertook was the development and implementation of an improved health and safety (H&S) training system. Figure 7.3 summarised the process of innovation, identifying both 'action' and 'reaction' forces.

Diagnosis

In July 2000 the principal of Contractor A identified that innovation was needed to 'drastically improve upon their training'. The firm had just obtained ISO 9002 and the perceived next step was to improve training in H&S. Contractor A has a good H&S record, but the principal wanted to make staff more aware of the H&S regulations, and improve skills and competency for the work carried out and, in doing this, improve productivity.

The need for innovation in H&S training was being driven by an increase in the turnover of site staff due to the firm's need to take on more staff because of its increased workload. Twelve months ago, the principal of

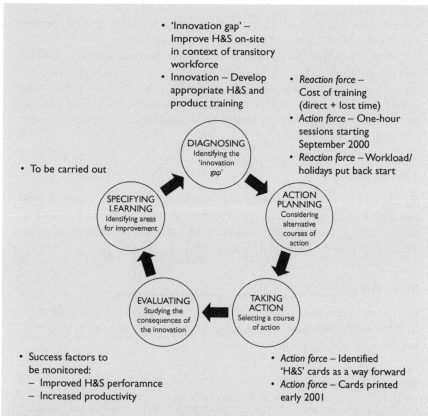

- 'Innovation gap' –
  Improve H&S on-site
  in context of transitory
  workforce
- Innovation – Develop
  appropriate H&S and
  product training

- *Reaction force –*
  Cost of training
  (direct + lost time)
- *Action force –* One-hour
  sessions starting
  September 2000
- *Reaction force –* Workload/
  holidays put back start

- To be carried out

DIAGNOSING
Identifying the
'innovation
gap'

SPECIFYING
LEARNING
Identifying areas
for improvement

ACTION
PLANNING
Considering
alternative
courses of
action

EVALUATING
Studying the
consequences of
the innovation

TAKING
ACTION
Selecting a course
of action

- Success factors to
  be monitored:
  – Improved H&S perforamnce
  – Increased productivity

- *Action force –* Identified
  'H&S' cards as a way forward
- *Action force –* Cards printed
  early 2001

Figure 7.3   The process of a H&S training innovation

Contractor A did not view staff turnover as a significant problem, with the majority of employees having been with the firm for some time. Due to the recent upturn in work, however, the firm needs to bring new people in, but they often do not stay long as they either are not good enough or want exorbitant wages. H&S regulations, however, demand that no matter how short a time employees stay with the firm, they still need to be aware of H&S.

Action planning

It was identified that training could be done either in-house or externally. Cost was perceived as a significant barrier or reactionary force to the innovation. External training was calculated to cost in the region of £900 to £1000 per head for a two-day course. Also labourers and trades people cost around £9 to £10 an hour; there were twenty employees, and one hour's training would probably cost two hours by getting them in.

Contractor A initially decided to introduce one-hour training sessions starting in August 2000. The sessions would be on basic H&S techniques. Heavy work commitments, coupled with the holiday period, united to cause the training sessions to be postponed until September 2000. In reality, the continuing workload pressures constantly pushed the development and start of the training back.

Taking action

By December, the focus and scale of the innovation were changed. The innovation was now to develop small H&S cards which would be issued to all site staff, giving both general and project specific H&S information. The H&S cards were being printed in early 2001.

Evaluating

Contractor A accepted that the evaluation of an improvement or otherwise in H&S would take time. The success factors would be improvements in performance and capabilities.

Specifying learning

Evaluation and learning will have to take place following the use of the H&S cards.

Summary

This chapter has presented key findings from the case studies on the process of innovation. It concluded that the process of innovation is not uniform, but rather it is characterised by an interplay between forces of 'action' and 'reaction' over time. Thus, the innovation can be successful, i.e. the 'action' forces, over time, overcome the 'reaction' forces, or unsuccessful, i.e. the 'reaction' forces are stronger than the 'action' forces.

The next chapter illustrates the discussion so far within the context of technology transfer in innovation within small construction firms.

Chapter 8

The role of technology transfer in innovation within small construction firms

Introduction

Technology transfer is widely considered to be a potentially powerful source of innovation which can provide construction firms with new technologies that can, where appropriate, transform and complement current technologies to create and sustain better levels of performance (for example, see Kogut and Zander, 1992; Nonaka and Takeuchi, 1995; Sexton et al., 1999). Technology transfer is viewed as the movement of knowledge and technology via some channel from one individual or firm to another (for example, see Devine et al., 1987; Gibson and Smilor, 1991; Inkpen and Dinur, 1998). Further, we take a broad view of technology, defining it as the know-how about the transformation of operational technologies and processes; material technologies; and knowledge technologies (for example, see Hickson et al., 1969; Wilson, 1986).

The construction industry delivers its product to its client base by way of a stream of generally single and unique projects. These projects typically draw together a significant number of diverse small and large construction firms into varying collaborations (for example, see Betts and Wood-Harper, 1994). The ambition to bring about the kind of step change improvements in construction industry performance called for by the Egan report (among others) must, by necessity, appropriately envision and engage large *and* small construction firms. Further, the scale of small firm activity in the UK construction industry is considerable, with, in 1999, 99 per cent of UK construction firms having between one and fifty-nine staff (DETR, 2000: Table 3.1), delivering some 52 per cent of the industry's workload in monetary terms (DETR, 2000: Table 3.3). Therefore any overall performance improvement of the industry through technology transfer is significantly influenced by the ability of small construction firms to absorb and use new technology.

The role of technology transfer in innovation in construction firms in general, and small firms in particular, is poorly understood, and there is a clear need to rectify this (for example, see Atkin, 1999; CRISP, 1999).

Key issues from the literature

Performance improvement based on technology absorbed into construction firms through technology transfer can and does occur successfully. Firms, however, need to understand and manage technology transfer activity to ensure consistent success. Sung and Gibson (2000) identified the following variables as affecting the degree of success in the process and results of technology transfer: person-to-person contacts; knowing whom to contact; variety of communication channels; set up transfer office or committee; a sense of common purpose; understanding of the nature of the business; attitude and values; increase in awareness of transfer; concreteness of knowledge and technology; establishment of a collaborative research programme; clear definition of transfer; and, provision of incentives for transfer and product champions. However, present construction industry technology transfer endeavours are being severely hampered by a lack of proper understanding of such technology transfer issues and their interrelationships to both company capabilities and processes, and the knowledge characteristics of the technologies being transferred, in particular (Barrett and Sexton, 1999):

- First, current approaches tend to view technology transfer as a mechanistic 'pick-and-mix' exercise – identifying new technologies, and trying to insert them in their existing form into (unsurprisingly) unreceptive construction firms.
- Second, current technology transfer mechanisms are not sufficiently informed by, or engaged with, company strategic direction and organisational capabilities and processes necessary to enable them to absorb technologies and to turn them into appropriate innovations. Experience from the manufacturing sector, for example, has stressed that the capacity of companies to understand and effectively use new technologies from external sources is heavily influenced by the level of prior-related knowledge and expertise (for example, see Adler and Shenhar, 1993).
- Finally, current technology transfer mechanisms do not fully appreciate that both the ability and motivation for construction firms to absorb and use new technologies is significantly influenced by the knowledge characteristics of the technologies. 'Hard' technologies which are characterised by explicit knowledge require very different diffusion mechanisms and organisational capabilities and processes from those required for 'soft' technologies, which are tacit in nature.

The implications of these barriers for technology transfer in small construction firms crystallises the systemic nature of technology transfer, and can be fruitfully viewed, as shown in Figure 8.1, as a 'technology transfer system' (Sexton et al., 1999).

- *Organisational direction and capability* – the motivation and ability of small construction firms to absorb and innovate from new technologies has to come

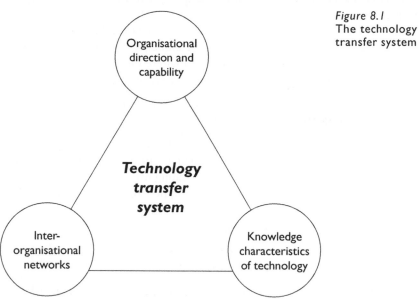

Figure 8.1
The technology
transfer system

from within the firm; through envisioning technology strategies and supporting organisational capabilities.

* *Inter-organisational networks* – small construction firms, along with all firms, do not operate in a vacuum; rather, they are situated in a number of fluctuating inter-organisational networks of varying complexity. Inter-organisational networks promote and facilitate the development and transfer of knowledge and resources needed to encourage learning and innovation in participating firms.

* *Knowledge characteristics of technology* – the extent to which new technology can be effectively absorbed by small construction firms is substantially influenced by the characteristics of the technology being transferred. Two characteristics are especially important. The first is the extent to which the knowledge embodied in the technology is explicit or tacit. Tacit knowledge is hard to formalise, making it difficult to communicate or share with others. Tacit knowledge involves intangible factors embedded in personal beliefs, experiences and values. Explicit knowledge is systematic and easily communicated in the form of hard data or codified procedures. The second characteristic is complexity. Whether based on explicit or tacit knowledge, some technologies are just more complex than others. The more complex a technology, the more difficult it is to unravel.

The argument is that technology transfer will be effective only if all three elements – organisational direction and capability, inter-organisational networks and the knowledge characteristics of technology – are appropriately focused and integrated to achieve a specific aim.

In summary, the literature stresses the important role of technology transfer in successful innovation and offers prescriptive guidance on how to manage the technology transfer process activity. Research within the construction industry, however, indicates significant barriers to effective technology transfer.

Key project results

The utility of the 'organisational factors of innovation' model (see Chapter 5) for understanding and managing technology transfer will be discussed by drawing upon a case study of one of the project collaborating firms (Sexton et al., 2006).

The case study describes Consultant B's absorption and use of an off-the-shelf computerised quantity surveying system. The case study will be structured around the 'organisational factors of innovation' model.

Given/interaction business environment

The project findings identifies clients as being the driving influence in the interaction environment. In common with all of the collaborating firms, Consultant B emphasised that: 'our strategy is very much driven from the outside by clients . . . it is difficult for a practice our size to be proactive'. This was the case with the trigger to adopt and use the new computerised quantity surveying system, with it being stressed that: 'our clients initially drove it . . . ten or twelve years ago we were working for [a UK water utility] and they insisted that all their bills be produced on a particular package'.

Business strategy and market positioning

The project findings suggest that owners of small construction firms view the increasing investment in information technology as an area of significant innovation activity in itself, and as a powerful enabler for innovation in enhancing the quality and efficiency of the services their firms offer. Consultant B believed that it needed to invest in specialist software to enable it to compete with its competitors. Having measured a job, the firm wanted to be able to produce a more flexible document:

> We wanted something that we could adapt and alter slightly, because, although it is based on a standard library, which is based on the standard method and measurement, which is the time-honoured way of producing a bill of quantities, the industry wanted things quicker, and we needed something that would allow us to make 'shortcuts' and wouldn't keep us in a 'straitjacket.' We've just done a job for which one document was required for three different

buildings. Basically once you've got one, it's very quick to do the other two. Much quicker than it would have been before; it is more efficient.

(Consultant B)

A further critical consideration in this sifting and evaluation process is the financial implications of a given technology transfer issue. The financial constraints faced by small construction firms affect the general capacity and capability for innovation. A partner of Consultant B argues:

small firms have a tight budget, so they don't have the people around to tackle a specific problem ... the cost of innovation is the short-term human involvement, and then having committed the capital to physically spend, you need some human time to make it work. The three go together. The big one though is the cash one.

This significant barrier to innovation is evident in Consultant B's future thinking for the computerised quantity surveying (QS) system:

we would like to digitise [our] QS system, enabling staff to quicken the speed at which they measure external areas of buildings. However, the idea is considered too expensive and would need guaranteed work to make it worthwhile.

The key argument being presented here is that owners of small construction firms need to be confident of the business benefit of absorbing and using a new technology before they will commit significant resources.

In summary, the business strategy and market positioning dimension to technology transfer is very much centred around an informal, intuitive process of identifying business needs and carrying out cost benefit analyses to determine optimal solutions. The owners are close enough to their firms' markets and capabilities to instinctively know what will work and what will not.

Technology

The computerised QS system chosen by Consultant B was a proven off-the-shelf solution. This was emphasised in the observation that: 'At the end of the day, it's a system that any QS could buy; it's not something that's specific to us'.

One of the partners of Consultant B stressed that although the software was, in itself, a piece of explicit, off-the-shelf technology, a significant amount of tacit knowledge had to be developed and shared before the technology could be absorbed into the firm and used. Indeed, it is acknowledged that 'we are fortunate, in many ways, that we have one guy who works here who lives and breathes computers' and that:

it is a fair comment to say that much of the knowledge needed [to use the software] is in people's heads . . . I think that what happens is that if someone notices that if you press Alt-B this happens, then the word gets around; but apart from that no, there's no conscious decision to disseminate the information.

In summary, small construction firms often lack the organisational capability and capacity to readily absorb and use technology requiring a high degree of tacit knowledge. Small construction firms focus on 'consumable' technology which can quickly and more easily be absorbed into the organisation by informal, mini-experimentation through 'learning-by-doing'.

Organisation of the work

The organisation of the technology transfer process involved Consultant B developing a relationship with the supplier of the quantity surveying software and the users of the software, i.e. the firm's staff.

The supplier relationship provided Consultant B with access to the technical expertise of the software house, and the experience of other firms using the package. The benefits and nature of this relationship were described as follows:

One of the benefits of the package is the support from the software people, so that if we say this didn't work well, or it would be handy to have this in here, they will look at it. They have workshops with various practices from around the country who use it.

The staff engaged in the technology transfer process at the piloting stage to ensure that the technology met the needs of the business and to nurture widespread ownership of the adoption of the new technology. The purpose and aim of the piloting stage were described as follows:

The job on which we trailed the packages was a very, very simple job, a little bungalow . . . I think it enabled us to choose the system; it showed us that (a) it worked, and (b) people were happy to use it, didn't find it too confusing and difficult.

The 'organisation of work' aspect then focused on developing supplier and external users' network to access expertise and experience, and to combine this with the recipient firm's own capabilities. The combining of network and firm knowledge was facilitated through internal piloting to enable safe mini-experimentation and staff empowerment of the new technology.

People

The staff of Consultant B needed the knowledge, skills and motivation to use the quantity surveying software properly. The knowledge and skills were developed in two ways. First, the software supplier provided three training days as part of the package. Second, as described in the 'Technology' section above, staff consolidated and developed their knowledge and skills through informal 'learning-by-doing'.

The ability of staff to use the new technology was not sufficient in itself; staff also had to be motivated to use it. This 'managing people through change' aspect was considered as core to the final success of the technology transfer. This imperative was captured in the following observation by one of the partners:

> I think we took people along with us when we were looking at it and making decisions; we didn't impose it. So people understood what we were trying to do and where we were trying to go. I don't think there was any doubt that we were going to do it. But because we wanted to keep the staff – we'd invested a lot of time and money in them – we wanted to take them along with us, and make sure they were happy.

In summary, the 'people' factor of innovation centres on developing the capability of staff to use new technologies. This principally takes the form of incoming expertise and experience from supplier and external users' networks (see 'Organisation of work' section) and 'learning-by-doing'. The motivation of staff to adopt new technology is important, with appropriate engagement and communication to effectively manage staff through change required.

Summary

This chapter has used the factors of innovation model set out in Chapter 5. The results reveal that small construction firms absorb and use technology which can contribute to the business in a quick, tangible fashion, and which can be dovetailed into organisational capabilities they already possess, or which can be readily acquired or 'borrowed' through their supplier and business networks. Any technology which is too far removed from this 'comfort zone', and which requires too much investment and contains too much risk, tends to be intuitively and swiftly sifted out. A safe evolution approach to innovation through technology transfer is taken as the way forward, not risky revolution.

The next, and final, chapter summarises the discussion set out in this book, and draws implications for innovation theory, policy makers and practitioners.

Chapter 9

Conclusion

Introduction

This book has presented key findings from investigative case studies to address the sparse literature on innovation in small construction firms, namely, the definition of innovation; the motivation to innovation; the organisational factors of innovation; the modes of innovation; and the process of innovation. This chapter summarises this research, and draws implications and recommendations. First, the contribution to innovation in small construction firms theory is presented. Second, recommendations are given for policy makers. Finally, recommendations are offered for small construction firm practitioners.

Contribution to innovation theory

The literature review set out in Chapter 2 explored the general and construction specific literature pertaining to innovation in small construction firms. The discussion was structured around the interrelated research gaps in the practice of innovation in small construction firms: innovation focus and outcomes; the context of innovation; organisational capabilities for innovation; and the process of innovation. These gaps formed the basis for a number of important research questions. These research questions are addressed below drawing upon the findings from case study investigation.

Focus and outcome of innovation

What is the generic strategic focus for innovation or definition of innovation for small construction firms? The literature review determined that in the general and construction literature there is an ongoing shift in the focus of innovation, from viewing innovation as an 'end' in itself, to innovation being a 'means' to achieve sustainable competitive advantage (see Chapter 2).

The case study findings confirm this shift, with the practitioners viewing successful innovation as:

the effective generation and implementation of a new idea, which enhances overall organisational performance.

This definition was developed originally by (and for) large construction firms (Barrett and Sexton, 1998), but is considered by the practitioners to be sufficiently inclusive to accurately define innovation in small construction firms. Applying this definition to small construction firms, the following assumptions are emphasised and illustrated:

- *Idea* – ideas are taken to mean the starting point for innovation (for example, see Thompson, 1965). Ideas can be administrative and technical in nature.
- *New* – not all ideas are recognised as innovations and it is accepted that newness is a key distinguishing feature (for example, see Zaltman et al., 1973). The idea has to be new only to a given firm, rather than new to the 'world'. Further, the newness aspect differentiates innovation from change. All innovation implies change, but not all change involves innovation. The examples of innovation offered by the firms tended to be the adoption of established ideas or technologies and/or their incremental adaptation. This is consistent with the view in the literature that small to medium sized firms rarely introduce fundamentally new products to their industry (Storey and Sykes, 1996), but are more likely to be involved in making incremental changes based on generic technologies than on more transformational changes (Rosenberg, 1992).
- *Effective generation and implementation* – innovation requires not only the generation of an idea (or transfer of a 'new' idea from outside the company), but also its successful implementation (for example, see Thompson, 1965). The implementation aspect differentiates innovation from invention (for example, see Monk, 1989).
- *Overall organisational performance* – innovation must improve organisational performance, either individually, or collectively through the supply chain (for example, see Kimberly, 1981).

What is the general motivation for small construction firms to innovate?

The literature review identified that firms needed to be appropriately motivated to innovate, but there is a general dearth of understanding of what the motivation to innovate is in small construction firms (see Chapter 2). The case study results provide new insights into this issue. The findings indicate that the motivation to innovate in small construction firms follows a fluid hierarchy of 'motivational needs', as shown in Figure 9.1 (see Chapter 4):

- *Survival* – small construction firms, owing to the type of markets they operate in and their lack of organisational resources and slack, concentrate foremost on project-based innovation focusing on survival.

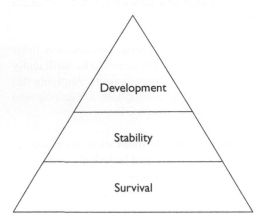

Figure 9.1   Hierarchy of motivational drivers for innovation

- *Stability* – it is only once survival has been confidently achieved that firms are sufficiently motivated to look towards consolidating and stabilising their market and/or resource position to ensure steady-state conditions over the medium term.
- *Development* – this stability provides the necessary motivation to exploit the prevailing stability and to develop and/or grow.

This type of hierarchy is consistent with arguments located in the stage theorist research literature. This literature describes stages 'as a configuration of organizational design variables representing a firm's response to the sets of dominant problems it faces at sequential times' (Kazanjian, 1988: 257). Flavoured by this type of thinking, Churchill and Lewis (1983), for example, describe five stages through which small firms pass: existence, survival, success, take-off and resource maturity. The case study findings, however, depart from this literature, by emphasising that survival, stability and development stages are not rigidly linear in progression, but cyclical in response to dynamic imbalances between shifting market-based and resource-based conditions (see Chapter 4). This dynamic and cyclical behaviour confirms that small construction firms remain more open to their external environments compared to large firms owing to their comparative lack of market and resource buffers.

Further, there is a tendency in the stage theory literature to assume that small firms strive for growth per se. The case study findings indicate that the motivation to innovate is not solely to grow, but can be directed at creating sustainable steady-state development.

It should be noted that although the case study findings stress the limited amount of organisational slack to enable innovation activity, it should not be presumed that more slack necessarily means better, large-scale innovation. As observed in the literature (for example, see Nohria and Gulati, 1996; Cheng and Kesner, 1997), there will come a point where additional resources channelled into innovation activity will generate ever decreasing returns, and will, in effect, be a waste of precious finite resources.

The principal implications of the case study findings on the motivation for small construction firms to innovate are threefold. First, small construction firms are not always motivated to innovate; when in 'survival' posture, firms will generally want to limit their exposure to the costs and risks of innovation as much as possible. Second, the hierarchy of motivational drivers for innovation (survival, stability and development) are dynamic and cyclical, not a linear progression. Third, not all small firms want to grow indefinitely in size; firm size will stabilise at a level which is compatible with the owner's aspirations.

What are common innovations outcomes in small construction firms?

The literature review catalogued a number of innovation outcomes for large and micro construction firms (see Chapter 2). The case study findings provide numerous and diverse examples of innovation outcomes: client relationship development innovation; organisation and management innovation at firm and project levels; technological innovation, etc. The principal outcome of innovation activity can be usefully grouped into two areas: improving the *effectiveness* of the firm, i.e. making sure that the firm is doing the right activities, and improving the *efficiency* of the firm, i.e. making sure that the firm's activities are done well.

Organisational capabilities for innovation

What are the key cognitive and organisational capabilities for innovation in small construction firms?

The literature review diagnosed two distinct, but complementary bundles of capabilities for innovation: cognitive (or thought) capabilities, and organisational (or action) capabilities. The literature was found, however, not to extend its consideration to an explicit understanding of what these capabilities were in small construction firms (see Chapter 2).

The case study findings produced a model of the organisational factors critical to successful innovation (see Figure 9.2). This model furnishes additional understanding of the required capabilities for innovation in small construction firms. The variables which make up the model are defined as follows:

- *Business strategy* is concerned with the overall purpose and longer term direction of the firm and its financial viability.
- *Market positioning* is the chosen (or emergent) orientation towards desired target markets for the purpose of achieving sustainable profitability.

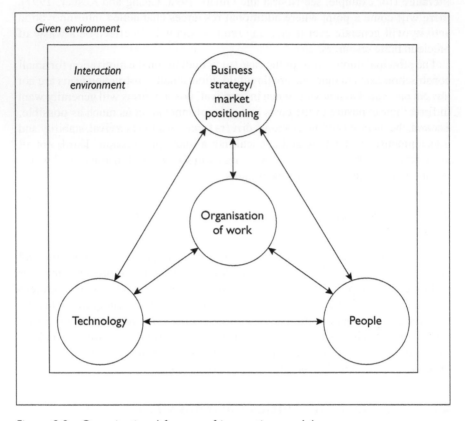

Figure 9.2 Organisational factors of innovation model

- *Technology* is the machines, tools and work routines used to transform material and information inputs (for example, labour, raw materials, components, capital) into outputs (for example, products and services).
- *People* are viewed as possessing knowledge, skills and motivation to perform a variety of tasks required to do the work of the firm.
- *Organisation of work* involves the creation and coordination of project teams and commercial networks both within the firm and across its business partners.
- *Interaction environment* is that part of the business environment which firms can interact with and influence.
- *Given environment* is that part of the business environment which firms are influenced by, but which they cannot influence themselves.

The model proposes that business strategy, market positioning, organisation of work, technology and people are the key organisational variables in understanding and improving innovation in small construction firms. The model emphasises and embraces both the holistic and systemic dimensions of innovation. The creation, management and exploitation of innovation involves consideration of not only the *content* of a chosen innovation, but also the management of the *process* of innovation and the *context* in which it occurs. The model considers two aspects of context: the inner and outer contexts of the firm. The inner context refers to the business strategy, market positioning, organisation of work, technology and people. The outer context refers to the 'given' and 'interaction' business environments. The process of innovation refers to the actions, reactions and interactions of, and between, the various organisational variables in the outer and inner contexts.

The organisational model argues that for practitioners to understand and organise innovation successfully, a broad, outwards looking approach is needed. It is a system, in the sense that all of the variables, shown in Figure 9.2, are interdependent and interrelated, as indicated by the double-headed arrows. For example, innovation in 'organisation of work' shapes and is shaped by all of the other variables. It depends on the skills of the people and technology availability to the firm. The system is open, in the sense that it depends on the external business environment for its survival.

How are these capabilities developed and used in innovation activity?

The literature review reported that the appropriate development and use of capabilities is critical to successful innovation, but that the literature did not specifically address how capabilities are developed and used in small construction firms. The case study findings reveal new understanding into how small construction firms develop and use the capabilities set out in the organisational model of innovation.

The *business strategy and market positioning* capability factors have three key implications for innovation in small construction firms. First, small construction firms are more exposed to the whims and movements of their business environments than large firms and, in necessary response, their business strategies tend to be more 'soft focus' and reactive in nature. Second, the greater vulnerability to the market amplifies the need for careful positioning in multiple markets and client relationship development in order to spread the risk of variable workflows in any one market. Third, the dominant role of the owner(s) of small firms allows quick decision-making and innovation activity to take place in response to rapidly shifting market conditions and client demands; in effect, to creating an agile firm. The very political strength exercised by owners, which can stimulate agility, can, if that individual lacks vision, bring about an adversely myopic view of the 'best way' for the firm to operate.

The *technology* capability factor has three key implications for innovation in small construction firms. First, information technology is an increasingly important *focus* for innovation in itself, and as an *enabler* for innovation. Second, 'soft' technologies, such as work routines, are particularly important for small construction firms. The required 'soft focus' and 'agility' of small firms to compete necessitates work routines which provide stability at the administrative core, and flexibility in project processes to adapt to rapidly changing market conditions and client needs. Third, knowledge management is seen as a way to transfer tacit knowledge located in individuals to company knowledge. This is particularly important for small construction firms, as often a significant proportion of their knowledge about clients and work activities are embodied in a small number of individuals. The concentration of knowledge in a few staff renders small firms especially vulnerable to key members of staff leaving the firm. Barrett and Ostergren (1991), for example, identify a number of adverse implications of the loss of critical staff for construction professional service firms, including leaving staff taking clients with them and eroding the goodwill of the firm.

The project findings on the *people* capability factor offer a number of important implications for innovation in small construction firms. First, the appropriate ability and motivation of staff is paramount for firms to create, manage and exploit innovation. Second, staff need to have a broad range of skills and experience to undertake multiple tasks. This flexibility is especially pertinent to small firms, who need to be 'agile' with limited, and often much stretched, staff resources.

The *organisation of work* capability factor identifies a number of key implications for innovation in small construction firms. First, there is a fairly uniform pattern of allocation of work through the hierarchy process which establishes reporting and communication structures at corporate and project levels. Second, the resourcing of projects and the organisation of project work is flexible in response to limited staff resources and the volatile and unpredictable characteristics of construction projects.

Context of innovation

What are the key precipitating events external and internal to small construction firms which trigger innovation activity?

The literature review observed that innovation activity in firms is triggered by 'precipitating' events in the firm's business environment and within the firm itself. It was identified that there is little understanding of what the key precipitating events are in small construction firms (see Chapter 2).

The case study findings developed our understanding of what these key precipitating events are. The principal contribution is that innovation activity in small construction firms is triggered predominantly by precipitating events in its external business environments, rather than within the firm itself. Innovation activity is stimulated in particular by changing client needs and unpredictable project-specific conditions. These triggers for innovation are predominantly filtered and prioritised by the owner(s) of the firm. This significant role of the interaction environment was discussed more fully in the next question.

What is the appropriate emphasis between market-based innovation and resource-based innovation in small construction firms, and what conditions dictate this emphasis?

The literature review developed the argument that there is an optimal balance of market-based or externally driven innovation and resource-based or internally driven innovation. The literature, however, was found to be deficient in what the appropriate emphasis in small construction firms is between market-based and resource-based innovation (see Chapter 2).

The case study findings identify two principal modes of innovation, shown in Figure 9.3 (see Chapter 6), which provides a better understanding of the shifting balance between market-based and resource-based innovation.

The two *modes of innovation* are shown in the centre portion of Figure 9.3. Mode 1 innovation focuses on progressing single project, cost-orientated relationships between the client and the firm – this mode of innovation is more driven by rapid change and uncertainty in the interaction environment, and the innovation is more market-based. Mode 2 innovation concentrates on progressing multiple project, value-orientated relationships between client and the firm – this mode of innovation is more aligned to improving the effectiveness of a firm's relationship with its clients and this mode of innovation stimulates an equal balance between market-based and resource-based innovation *market*, and enhancing the effectiveness of its *resources*.

The right-hand side of Figure 9.3 reinforces the notion that the mode of innovation is substantially determined by the nature of the *interaction environment*: an enabling interaction environment encourages Mode 2 innovation, a constraining environment is conducive to Mode 1 innovation. An enabling interaction environment is one which the firm can influence to a significant extent, *enabling* the firm to innovate within a longer term and more secure context. A constraining interaction environment is one which a small construction firm can influence only to a limited extent, *constraining* the firm to innovation activity undertaken within a shorter and more insecure context.

The left-hand side of Figure 9.3 identifies which factors of the organisational model are the primary focuses of (and levers for) innovation activity: Mode 2 innovation involves innovation in the business strategy and market positioning variables which, in turn, will have implication for the remaining variables; Mode 1 innovation is where the business strategy and market positioning variables are relatively fixed, and the focus of the activity is in the organisation of work, technology and people variables. Mode 1 and Mode 2 innovation presents a continuum, rather than a choice of two discrete types of innovation activity, i.e. 'hybrid' modes of innovation can be located between Mode 1 and Mode 2.

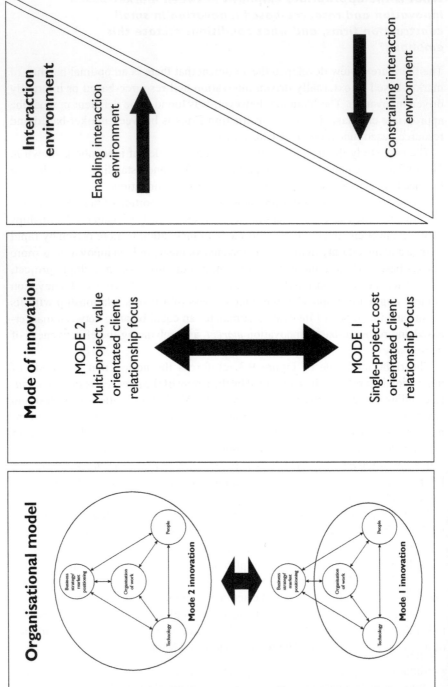

Figure 9.3 Mode 1 and Mode 2 innovation

Process of innovation

Are the processes of innovation in small construction firms rational and/or behavioural in nature?

The literature exploring the process of innovation has developed two schools of thought: first, that the process of innovation is rational in character; second, that the process of innovation is behavioural in nature (see Chapter 2).

The case study findings reveal that the process of innovation is very much behavioural in nature, and can be loosely conceptualised as an adapted, ongoing action research cycle (see Chapter 7). This process of innovation is shown in Figure 9.4. As we saw in Chapter 7, the process has five parts:

1   *Diagnosis*, where the issue (be it an opportunity or problem) forming the focus of the innovation activity is identified, and information is collected for a more detailed diagnosis.
2   *Action plan* is prepared after the diagnosis, where possible ways to progress the innovation are developed, and from which an agreed plan of action emerges. This provides the basis for the next step.
3   *Taking action* is when the idea is put into practice.
4   *Evaluation* next takes place to determine whether the innovation has been a success or not.
5   *Specific learning* is then undertaken, where innovation is reassessed, areas for improvement identified, and the process begins another cycle.

The cycle starts with sensing an opportunity or need to innovate in response to market, project and/or client conditions. These triggers for innovation are predominantly filtered and prioritised by the owner(s) of the firm.

Innovation activity does not take place in five sequential stages; rather, as depicted in the outer ring of Figure 9.4, the cycle can take place at each stage of the 'overall' process. At the diagnosis stage, for example, the practitioner might well go through periods of evaluation and reflection to confirm that the 'innovation gap' identified is appropriate. This 'overall' process continues until the innovation is either successful, or it is decided that the innovation is not appropriate, and should not be continued.

The process of innovation is not uniform, however; rather, as shown in Figure 9.5, the process is characterised by an interplay between forces of 'action' and 'reaction' over time which progresses or inhibits the closing of the 'innovation gap' between the current level of performance and a desired level of performance.

There are a myriad of potential action and reaction forces in each of the organisational innovation variables, such as, strong senior management support for the innovation (action), resistance to change from staff (reaction), allocation of capital to purchase needed technology (action), and lack of appropriate work routines to coordinate and channel the innovation activity (reaction). Figure 9.5 presents a situation where the innovation is successful, i.e. the 'action' forces, over time, have

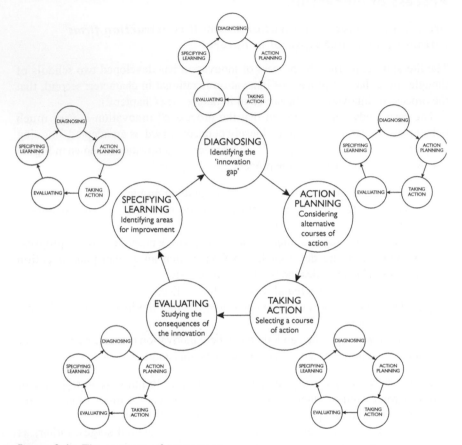

Figure 9.4   The process of innovation

overcome the 'reaction' forces. It is just as feasible, of course, for the 'reaction' forces to be stronger than the 'action' forces, and for the innovation to fail.

Implications for policy makers

The case study findings have identified two key implications for policy makers: to understand and accommodate the different needs of large and small construction firms, and to appreciate the value of small construction firms being engaged with collaborative research. Each of these implications will be discussed in turn.

Appropriate, targeted policies and guidance

The case study findings show that small construction firms have their own distinctive characteristics and needs which are significantly different from those

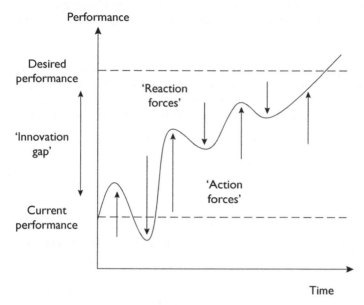

Figure 9.5   The innovation process

of large construction firms – small construction firms are not merely large construction firms scaled down. Policies to improve performance of construction firms need to understand and actively manage these differences; policies which are appropriate for large construction firms are not necessarily appropriate for small construction firms, and vice versa. In particular, the following should be noted when developing policy:

- Small construction firms ideally need an 'enabling' interaction environment to innovate within a longer term and more secure context. Policy needs to show the shared benefits to clients and large construction firms of creating and supporting this type of environment for small construction firms.
- Initiatives such as Construction Excellence are suitable vehicles to demonstrate this, developing tailor-made, and thus more targeted, guidance, best practice and case studies for both clients, and large and small construction firms.
- Small construction firms have limited staff, money and time, all of which are under greater pressure compared to large construction firms. Policy needs to appreciate this 'leanness' and promote initiatives which encourage small construction firms to leverage their existing resources, rather than initiatives which need additional resources. This issue is closely connected with the appropriateness of policies and guidance for small construction firms discussed above. The challenge is for policies to address appropriate issues to stimulate

interest in small firm practitioners, and to be in a format and language which is conducive to being easily understood and used from a small construction firm perspective.

The professional institutions, such as the Royal Institute of Chartered Surveyors and the National Federation of Builders, have a key role in this respect. Professional institutions are in an ideal position to act as knowledge brokers, to filter, package and disseminate new ideas and best practice which are appropriate for small construction firms' specific needs.

Engaging small construction firms in collaborative research

Collaborative research within the construction industry between construction firms and universities and similar research bodies has consistently demonstrated its ability to enhance the performance of construction firms, to better satisfy clients' and society needs and to improve the quality of the built environment (for example, see Lissenburgh and Harding, 2000). There is a growing appreciation, however, that the generation of construction research and innovation must meet the needs of large *and* small construction firms within the industry, and that this must be coupled with *appropriate* dissemination mechanisms to develop awareness of the research findings available, and to encourage and facilitate their appropriate exploitation.

The case study investigation has demonstrated that small construction firms are willing and able to engage in collaborative research, and have benefited from the experience. This view is evidenced by the seven firms in this project which invested time and effort over an eighteen-month period, and which tangibly benefited from the sharing of knowledge and experience.

Research funding bodies should be encouraged by this example, and continue to develop research initiatives which actively encourage small construction firm participation by focusing on small firm opportunities and problems. In particular, programmes which address firm-specific issues and which contribute to direct costs, such as the Knowledge Transfer Partnership scheme, should be expanded. In much the same vein, the resource limitations of small construction firms should be recognised in collaborative research which demands matched funding from industry.

Recommendations for practitioners

The case study investigation offers four key ways of enabling practitioners to better understand, manage and exploit innovation activity.

First, the owner(s) of small construction firms have the necessary power to ensure quick decision-making and innovation activity to take place in response to rapidly shifting market conditions and client demands; in effect, creating an agile firm. These triggers for innovation are predominantly filtered and prioritised by the owner(s) of the firm. The dominant role of the owner, however, can constrain innovation activity if the owner does not have the necessary *vision* and *systemic thinking* when diagnosing and progressing innovation activity. Innovation in one part of the business often has significant implications for other parts of the business which need to be considered and brought together in an integrated way.

Second, to assist in this the organisational model of innovation identifies the factors critical to successful innovation: 'business strategy', 'market positioning', 'organisation of work', 'technology' and 'people' (see Chapter 5). The model provides a framework or checklist to help the practitioner identify what action has to be taken to progress an innovation in a systemic, integrated way. The type of innovation undertaken and the different organisational factors which are brought into play depend to a significant extent on the characteristics of the interaction environment in which the firm is operating.

Third, to aid the practitioner in knowing what type of innovation activity to pursue in any given interaction environment the *modes of innovation model* identifies two principal modes of innovation: Mode 1, single-project, cost-orientated client relationships, and Mode 2, multi-project, value-orientated client relationships (see Chapter 6). The model provides the practitioner with an understanding that the mode of innovation is substantially determined by whether the interaction environment is enabling or constraining. Mode 1 innovation is principally concerned with innovation in the organisation of work, technology and people variables of the organisational model of innovation, with the business strategy and market positioning variables being relatively fixed. Mode 2 innovation is focused on innovation in the business strategy and market positioning variables of the organisational model of innovation, with associated implications for the remaining variables of organisation of work, technology and people.

The key implication for small construction firms is that they should not 'flip' from Mode 1 innovation to Mode 2 innovation. Small construction firms need to incrementally nurture, or identify and move into, supportive enabling interaction environments. This is achieved through careful and integrated consideration and development of all the variables in the organisation model of innovation in an integrated fashion. Whatever the mode of innovation, innovation does not just happen in a small construction firm; rather there is a process of innovation.

Finally, to better understand this process the *process of innovation model* offers a cyclical process of diagnosing, action planning, taking action, evaluating and specifying learning (see Chapter 7). The cycle starts with sensing an opportunity or need to innovate in response to market, project and/or client conditions. These triggers for innovation are predominantly filtered and prioritised by the owner(s) of the firm. The *action and reaction in innovation model* gives the practitioner confidence that the innovation process is often not an orderly, neat process, but a process that is subject to peaks and troughs as the progress of the innovation competes with day-to-day variability of workload and the often acute pressures on finite staff and financial resources.

Envoi

The practice-focused guidance given in this book, even though drawn from case study research material and practitioner debate, is not considered *the* right way forward. Each small construction firm will have its own unique needs and characteristics that drive and shape what is appropriate innovation and what is the best way to understand and manage innovation activity. But on the basis that the guidance which is offered worked well for a range of different collaborating firms in practice, it is hoped that it has real potential to help small construction firms to develop their own distinctive, successful approach to innovation.

References

Abdel-Razek, R. and McCaffer, R. (1987) 'A Change in the UK Construction Industry Structure: Implications for Estimating', *Construction Management and Economics*, 5: 227–242.

Adler, P.S. and Shenhar, A. (1993) 'Adapting your Technology Base: The Organizational Challenge', *Sloan Management Review*, 32 (1): 25–37.

Amabile, T.M., Conti, R., Coon, H., Lazenby, J. and Herron, M. (1996) 'Assessing the Work Environment for Creativity', *Academy of Management Journal*, 39: 1154–1184.

Andreu, R. and Ciborra, C. (1996) 'Organisational Learning and Core Capabilities Development: The Role of IT', *Journal of Strategic Information Systems*, 5: 111–127.

Argyris, C. and Schön, D. (1978) *Organizational Learning: A Theory-in-Action Perspective*, Addison-Wesley, Reading, MA.

Artto, K., Heinonen, R., Arenius, M., Kovanen, V. and Nyberg, T. (1998) *Global Project Business and the Dynamics of Change*, Technology Development Centre/Project Management Association, Helsinki, Finland.

Atkin, B. (1999) *Innovation in the Construction Sector*, European Council for Construction Research, Development and Innovation, Brussels.

Atkin, B. and Pothecary, E. (1994) *Building Futures: A Report on the Future Organisation of the Building Process*, University of Reading, Reading, UK.

Ball, M. (1988) *Rebuilding Construction: Economic Change in the British Construction Industry*, Routledge, London.

Banfield, P., Jennings, P.L. and Beaver, G. (1996) 'Competence-based Training for Small Firms: An Expensive Failure?', *Long Range Planning*, 29 (1): 94–102.

Banwell, H. (1964) *Report of the Committee on the Placing and Management of Contracts for Building and Civil Engineering Work*, HMSO, London.

Barnett, E. and Storey, J. (2000) 'Managers' Accounts of Innovation Processes in SMEs', *Journal of Small Business and Enterprise Development*, 7 (4): 315–324.

Barney, J.B. (1991) 'Firm Resources and Sustained Competitive Advantage', *Journal of Management*, 17: 99–120.

Barrett, P. (ed.) (1995) *Facilities Management: Towards Best Practice*, Blackwells Scientific, Oxford.

Barrett, P. and Ostergren, K. (1991) 'The Value of Keypersons in Professional Firms', in P. Barrett and R. Males (eds) *Practice Management, New Perspectives for the Construction Professional*, E & FN Spon, London.

Barrett, P. and Sexton, M.G. (1998) *Integrating to Innovate: Report for the Construction Industry Council*, DETR/CIC, London.

Barrett, P. and Sexton, M.G. (1999) 'The Transmission of "Out-of-industry" Knowledge into Construction Industry Wisdom', Linking Construction Research and Innovation in Other Sectors, Construction Research and Innovation Strategy Panel, London, 24 June.

Barrett, P. and Stanley, C. (1999) *Better Construction Briefing*, Blackwell Science, Oxford.

Betts, M. (ed.) (1999) *Strategic Management of I.T. in Construction*, Blackwell Science, Oxford.

Betts, M. and Wood-Harper, T. (1994) 'Reengineering Construction: A New Management Research Agenda', *Construction Management and Economics*, 12: 551–556.

Bijker, W.E., Hughes, T.P. and Pinch, T.J. (1987) *The Social Construction of Technological Systems*, MIT Press, Cambridge, MA.

Blatt, R. (1993) *Young Company Study: 1989–1992*, Ministry of Economic Development and Trade, Government of Ontario, Toronto, Canada.

Bolton, J. (1971) *Small Firms – Report of the Committee of Inquiry on Small Firms*, HMSO, London.

Bowley, M. (1960) *Innovation in Building Materials*, Duckworth, London.

Bracker, J.S. and Pearson, J.N. (1986) 'Planning and Financial Performance of Small Mature Firms', *Strategic Management Journal*, 7: 503–522.

Briscoe, G., Dainty, A.R.J. and Millet, S.J. (2000) 'The Impact of the Tax System on Self-employment in the British Construction Industry', *International Journal of Manpower*, 21 (8): 596–613.

Brouseau, E. and Rallet, A. (1995) 'Efficacité et inefficacité de l'organisation du bâtiment: Une interprétation en termes de trajectoire organisationalle', *Revue d'Economie Industrielle*, 74 (4): 9–30.

Burgelman, R., Maidique, M. and Wheelwright, S. (1996) *Strategic Management of Technology and Innovation*, Irwin, Homewood, IL.

Burns, T. and Stalker, G.M. (1961) *The Management of Innovation*, Tavistock, London.

Capaldo, G., Corti, E. and Greco, O. (1997) 'A Coordinated Network of Different Actors to Offer Innovation Services to Develop Local SMEs Inside Areas with a Delay of Development', *Proceedings of the ERSA Conference*, Rome, 26–29 August.

Carty, G. (1995) 'Construction', *Journal of Construction Engineering and Management*, 121 (3): 319–328.

CERF (Civil Engineering Research Foundation) (2000) *Guidelines for Moving Innovations into Practice*, Working Draft Guidelines for the CERF International Symposium and Innovative Technology Trade Show 2000, CERF, Washington, DC.

Chaston, I., Badger, B. and Sadler-Smith, E. (1999) 'Organisational Learning: Research Issues and Application in SME Sector Firms', *International Journal of Entrepreneurial Behaviour and Research*, 5 (4): 191–203.

Cheng, J.J.C. and Kesner, I.F. (1997) 'Organizational Slack and Response to Environmental Shifts: The Impact of Resource Allocation Patterns', *Journal of Management*, 23 (1): 1–18.

Child, J. (1997) 'Strategic Choice in the Analysis of Action, Structure, Organizations and Environment: Retrospect and Prospect', *Organization Studies*, 18 (1): 43–76.

Chisnell, P.M. (1995) *Strategic Business Market*, Prentice-Hall, New York.

Christensen, C.M. (1997) *The Innovator's Dilemma: When New Technology Cause Great Firms to Fail*, Harvard Business School Press, Boston, MA.

Churchill, N.C. and Lewis, V.L. (1983) 'The Five Stages of Small Business Growth', *Harvard Business Review*, 61 (3): 30–50.

CIB Task Group 35 (2000) *Innovation in the British Construction Industry: The Role of Public Policy Instruments*, University College London, London.

Cohen, W.M. and Levinthal, D.A. (1990) 'Absorptive Capacity: A New Perspective on Learning and Innovation', *Administrative Science Quarterly*, 35: 128–152.

Construction Productivity Network (1997) *Human Resources for Construction Innovation: Report of IMI International Workshop*, University of Reading, Reading, UK, 19–20 May.

Cooper, R., Hinks, J., Aouad, G., Kagioglou, M., Sheath, D. and Sexton, M.G. (1998) *A Generic Guide to the Design and Construction Process Protocol*, University of Salford, Salford.

CRISP (1999) *Workshop on Linking Construction Research and Innovation to Research and Innovation to Research and Innovation in Other Sectors*, CRISP, London, 24 June.

Cross, R. (1998) 'Managing for Knowledge: Managing for Growth', *Knowledge Management*, 1 (3): 9–13.

Curran, J. and Blackburn, R.A. (2001) *Researching the Small Enterprise*, Sage, London.

Cyert, R. and March, J. (1963) *A Behavioral Theory of the Firm*, Blackwell, Oxford.

Davies, S. (1979) *The Process of Diffusion Innovations*, Cambridge University Press, Cambridge.

De Koning, A. and Snijders, J. (1992) 'Policy on Small- and Medium-sized Enterprises in Countries of the European Community', *International Small Business Journal*, 10 (3): 25–39.

Department of the Environment, Transport and the Regions (DETR) (1998) *Rethinking Construction* (Egan Report), DETR, London.

Department of the Environment, Transport and the Regions (DETR) (2000) *Construction Statistics Annual: 2000 Edition*, DETR, London.

Department of Trade and Industry (DTI) (2003a) *Construction Statistics Annual: 2003 Edition*, HMSO, London.

Department of Trade and Industry (DTI) (2003b) *UK 2001 Census Statistics*, HMSO, London.

Devine, M.D., James, T.E. and Adams, I.T. (1987) 'Government Supported Industry Research Centres: Issues for Successful Technology Transfer', *Journal of Technology Transfer*, 12 (1): 27–38.

Dodgson, M. and Rothwell, R. (1991) 'Technology Strategies in Small Firms', *Journal of General Management*, 17 (1): 45–55.

Donaldson, L. (1988) 'In Successful Defence of Organization Theory: A Routing of the Critics', *Organization Studies*, 9 (1): 28–32.

Dosi, G. (1982) 'Technological Paradigms and Technological Trajectories: A Suggested Interpretation of the Determinants and Directions of Technical Change', *Research Policy*, 11: 146–162.

Dosi, G. (1984) *Technical Change and Industrial Transformation*, Macmillan, London.

Doty, D.H., Glick, H.W. and Huber, P.G. (1993) 'Fit, Equifinality, and Organizational Effectiveness: A Test of Two Configurational Theories', *Academy of Management Journal*, 36 (6): 1196–1250.

Drucker, P.F. (1986) *Innovation and Entrepreneurship*, Pan, London.

Egan, J. (1998) *Rethinking Construction*, Report from the Construction Task Force, Department of the Environment, Transport and Regions (DETR), London.

Emmerson, H. (1962) *Studies of Problems before the Construction Industries*, HMSO, London.

Emmitt, S. (2002) *Architectural Technology*, Blackwell Science, London.

European Commission (1995) *Green Paper on Innovation*, European Commission, Brussels.

European Commission (2003) *2003 Observatory of European SMEs*, European Commission, Luxembourg.

Fairclough, J. (2002) *Rethinking Construction Innovation and Research: A Review of Government R&D Policies and Practices*, Department for Transport, Local Government and the Regions (DTLR), HMSO, London.

FIEC (European Construction Industry Federation) (2003) *Construction in Europe: Key Figures*, Leaflet, August.

Finkelstein, S. and Hambrick, D.C. (1990) 'Top-management Tenure and Organizational Outcomes: The Moderating Role of Management Discretion', *Administrative Science Quarterly*, 35: 484–503.

Freeman, C. (1989) *The Economics of Industrial Innovation*, Pinter, London.

Gale, A.W. and Fellows, R.F. (1990) 'Challenge and Innovation: The Challenge to the Construction Industry', *Construction Management and Economics*, 8 (4): 431–436.

Gann, D.M. (1998) 'Do Regulations Encourage Innovation? The Case of Energy Efficiency in Housing', *Building Research and Information*, 26 (5): 280–296.

Gann, D.M. (2000) *Building Innovation: Complex Constructs in a Changing World*, Thomas Telford, London.

Gann, D.M. and Salter, A.J. (2000) 'Innovation in Project-based, Service-enhanced Firms: The Construction of Complex Products and Systems', *Research Policy*, 29 (7–8): 955–972.

Gibson, D. and Smilor, R. (1991) 'Key Variables in Technology Transfer: A Field-study Based Empirical Analysis', *Journal of Engineering and Technology Management*, 8: 287–312.

Grant, R.M. (1997) *Contemporary Strategic Analysis: Concepts, Techniques, Applications*, 3rd edition, Blackwell, Oxford.

Gray, C. (1998) *Enterprise and Culture*, Routledge, London.

Hadjimanolis, A. (2000) 'A Resource-based View of Innovativeness in Small Firms', *Technology Analysis and Strategic Management*, 12 (2): 263–281.

Hamel, G. and Prahalad, C. (1994) *Competing for the Future*, Harvard Business School Press, Boston, MA.

Herzberg, F., Mausner, B., and Snyderman, B. B. (1959) *The Motivation to Work*, Wiley, New York.

Hewitt-Dundas, N. and Roper, S. (2000) *Strategic Re-engineering: Small Firms' Tactics in a Mature Industry*, Northern Ireland Economic Research Centre Working Paper Series no. 49, Northern Ireland Economic Research Centre, Belfast, Northern Ireland.

Hickson, D.J., Pugh, D.S. and Pheysey, D.C. (1969) 'Operations Technology and Organization Structure: An Empirical Reappraisal', *Administrative Science Quarterly*, 14: 378–379.

Howell, J.M. and Higgins, C.A. (1990) 'Champions of Technological Innovation', *Administrative Science Quarterly*, 35: 317–341.

Hoxley, M. (1993) 'Obtaining and Retaining Clients: A Study of Service Quality and the Referral Systems of U.K. Building Surveying Practices', unpublished MPhil thesis, University of Salford, Salford, UK.

Inkpen, A.C. and Dinur, A. (1998) 'Knowledge Processes and the International Joint Ventures', *Organization Science*, 9 (4): 454–468.

Jenson, M. (1986) 'Agency Costs of Free Cash Flow, Corporate Finance and Takeovers', *American Economic Review*, 76: 323–329.

Julien, P.A. (1996) 'Information Control: A Key Factor in Small Business Development', *Conference Proceedings of the Forty-first ICSB World Conference*, Stockholm, 17–19 June.

Kanter, R.M. (1984) *The Change Masters: Innovations for Productivity in the American Corporation*, Simon & Schuster, New York.

Kast, F.E. and Rosenzweig, J.E. (1973) *Contingency Views of Organization and Management*, Science Research Associates, Palo Alto, CA.

Kazanjian, R.K. (1988) 'Relation of Dominant Problems to Stages of Growth Technology-based New Ventures', *Academy of Management Journal*, 31 (2): 257–279.

Kimberly, J.R. (1981) 'Managerial Innovation', in P.C. Nystrom and W.H. Starbuck (eds) *Handbook of Organizational Design, Volume 1: Adapting Organizations to their Environments*, Oxford University Press, New York.

Kogut, B. and Zander, U. (1992) 'Knowledge of the Firm, Combinative Capabilities, and the Replication of Technology', *Organization Science*, 3: 383–397.

Kolb, D. (1984) *Experiential Learning*, Prentice-Hall, Englewood Cliffs, NJ.

Kotler, P. (1980) *Marketing Management: Analysis, Planning and Control*, Prentice-Hall, Englewood Cliffs, NJ.

Langford, D.A. and Male, S. (1992) *Strategic Management in Construction*, Gower, Aldershot.

Latham, M. (1994) *Constructing the Team: Joint Review of Procurement and Contractual Arrangements in the UK Construction Industry*, Department of the Environment, HMSO, London.

Laudau, R. and Rosenberg, N. (eds) (1986) *The Positive Sum Strategy*, National Academy Press, Washington, DC.

Lawler, E.E. (1973) *Motivation in Work Organizations*, Brooks / Cole, Monterey, CA.

Lawrence, P.R. and Lorsch, J.W. (1967) *Organization and Environment: Managing Differentiation and Integration*, Harvard University Press, Cambridge, MA.

Leonard, D. (1995) *Wellsprings of Knowledge: Building and Sustaining the Sources of Innovation*, Harvard Business School Press, Boston, MA.

Lissenburgh, S. and Harding, R. (2000) *Knowledge Links: Innovation in University/ Business Partnerships*, Institute for Public Policy Research, Southampton.

Louis, M.R. and Sutton, R.I. (1991) 'Switching Cognitive Gears: From Habits of Mind to Active Thinking', *Human Relations*, 44: 55–76.

Maidique, M.A. (1980) 'Entrepreneurs, Champions, and Technological Innovation', *Sloan Management Review*, 21 (2): 59–76.

Maijoor, S. and van Witteloostuijn, A. (1996) 'An Empirical Test of the Resource-based Theory: Strategic Regulation in the Dutch Audit Industry', *Strategic Management Journal*, 17: 549–569.

Maister, D.H. (1993) *Managing the Professional Service Firm*, Free Press, New York.

Manseau, A. and Seaden, G. (2001) *Innovation in Construction: An International Review of Public Policies*, Spon Press, London.

Mansfield, E., Rapoport, J., Schnee, J., Wagner, S. and Hamburger, M. (1971) *Research and Innovation in the Modern Corporation*, Norton, New York.

Mason, C., McNally, K. and Harrison, R. (1996) 'Sources of Equity Capital for Small Growing Firms: Acost's "Enterprise Challenge" Revisited', in R. Oakey (ed.) *New Technology-based Firms in the 1990s*, Volume 2, Chapman, London.

Miles, R.E. and Snow, C.C. (1994) *Fit, Failure, and the Hall of Fame: How Companies Succeed or Fail*, Free Press, New York.

Miller, D. and Friesen, P.H. (1984) *Organizations: A Quantum View*, Prentice-Hall, Englewood Cliffs, NJ.

Miller, D. and Toulouse, J.M. (1986) 'Chief Executive Personality and Corporate Strategy and Structure in Small Firms', *Management Science*, 32 (11): 1389–1409.

Miozzo, M. and Ivory, C. (1998) *Innovation in Construction: A Case Study of Small and Medium-sized Construction Firms in the North West of England*, Manchester School of Management, UMIST, Manchester, UK.

Mitchell, G. and Hamilton, W. (1988) 'Managing R&D as a Strategic Option', *Research Technology Management*, 31: 15–22.

Mohr, L.B. (1969) 'Determinants of Innovation in Organizations', *American Political Science Review*, 63: 111–126.

Mohsini, R.A. and Davidon, C.H. (1992) 'Detriments of Performance in the Traditional Building Process', *Journal of Construction Management and Economics*, 10: 343–359.

Monk, P. (1989) *Technological Change in the Information Economy*, Pinter, London.

Motawa, I.A., Price, A.D.F. and Sher, W. (1999) 'Scenario Planning for Implementing Construction Innovation', *Proceedings of the RICS Construction and Building Research Conference (COBRA 1999) – The Challenge of Change: Construction and Building for the New Millennium*, University of Salford, Salford, UK.

Mullins, L.J. (1999) *Management and Organisational Behaviour*, 5th edition, Pearson Education, Harlow.

Nam, C.B. and Tatum, C.B. (1997) 'Leaders and Champions for Construction Innovation', *Construction Management and Economics*, 15: 257–270.

Nelson, R.R. and Winter, S.G. (1982) *An Evolutionary Theory of Economic Change*, Harvard University Press, Cambridge, MA.

Nohria, N. and Gulati, R. (1996) 'Is Slack Good or Bad for Innovation?', *Academy of Management Journal*, 39 (5): 1245–1264.

Nonaka, I. and Takeuchi, H.T. (1995) *The Knowledge-creating Company: How Japanese Companies Create the Dynamics of Innovation*, Oxford University Press, New York.

Nooteboom, B. (1994) 'Innovation and Diffusion in Small Firms: Theory and Evidence', *Small Business Economics*, 6: 327–347.

Nordhaug, O. (1993) *Human Capital in Organizations*, Scandinavian University Press, Oslo.

OECD (Organisation for Economic Co-operation and Development) (1991) *Proposed Guidelines for Collecting and Interpreting Technological Innovation Data (Oslo Manual)*, OECD, Oslo, 17 September.

O'Farrell, P.N. and Hitchens, D.M. (1988) 'Alternative Theories of Small-firm Growth: A Critical Review', *Environment and Planning A*, 20 (10): 1365–1383.

Parker, J. (1978) *The Economics of Innovation: The National and Multinational Enterprise in Technological Change*, Longman, London.

Pearce, D. (2003) *The Social and Economic Value of Construction: The Construction Industry's Contribution to Sustainable Development* (Pearce Report), nCRISP, London.

Peters, T.J. and Waterman, R.H. (1982) *In Search of Excellence: Lessons from America's Best Run Companies*, Harper & Row, New York.

Polanyi, M. (1967) *The Tacit Dimension*, Doubleday Anchor, New York.

Porter, M.E. (1980) *Competitive Strategy: Techniques for Analyzing Industries and Competitors*, Free Press, New York.

Porter, M.E. (1985) *Competitive Advantage: Creating and Sustaining Superior Performance*, Free Press, New York.

Porter, M.E. (1991) 'Know Your Place', *Inc. Magazine*, September: 90–93.

Powell, J. (1995) *Towards a New Construction Culture*, Chartered Institute of Building, Ascot, UK.

Prahalad, C.K. and Hamel, G. (1990) 'The Core Competence of the Corporation', *Harvard Business Review*, 68 (3): 71–91.

Pries, F. and Janszen, F. (1995) 'Innovation in the Construction Industry: The Dominant Role of the Environment', *Construction Management and Economics*, 13: 259–270.

Quinn, J.B. (1985) 'Managing Innovation: Controlled Chaos', *Harvard Business Review*, 63 (3): 73–84.

Reichstein, T., Salter, A.J. and Gann, D.M. (2005) 'Last among Equals: A Comparison of Innovation in Construction, Services and Manufacturing in the UK', *Construction Management and Economics*, 23 (6): 631–644.

Reid, G.C. and Jacobsen, L.R. (1988) *The Small Entrepreneurial Firm*, Aberdeen University Press, Aberdeen.

Rosenberg, N. (1982) *Inside the Black Box: Technology and Economics*, Cambridge University Press, Cambridge.

Rosenberg, N. (1992) 'Science and Technology in the Twentieth Century', in G. Dosi (ed.) *Technology and Enterprise in Historical Perspective*, Clarendon Press, Oxford.

Rothwell, R. (1989) 'Small Firms, Innovation and Industrial Change', *Small Business Economics*, 1: 51–64.

Rothwell, R. (1991) 'External Networking and Innovation in Small and Medium Size Manufacturing Firms in Europe', *Technovation*, 11 (2): 93–112.

Rothwell, R. and Dodgson, M. (1994) 'Innovation and Firm Size', in M. Dodgson and R. Rothwell (eds) *The Handbook of Industrial Innovation*, Edward Elgar, Aldershot, UK.

Rothwell, R. and Zegveld, W. (1985) *Reindustrialization and Technology*, Longman, London.

Sahal, D. (1981) 'Alternative Conceptions of Technology', *Research Policy*, 10: 2–24.

Schot, J.W. (1992) 'Constructive Technology Assessment and Technology Dynamics: The Case of Clean Technologies', *Science, Technology and Human Values*, 17 (1): 36–55.

Sexton, M. and Barrett, P. (2003a) 'A Literature Synthesis of Innovation in Small Construction Firms: Insights, Ambiguities and Questions', *Construction Management and Economics*, 21 (6): 613–622.

Sexton, M. and Barrett, P. (2003b) 'Appropriate Innovation in Small Construction Firms', *Construction Management and Economics*, 21 (6): 623–633.

Sexton, M. and Barrett, P. (2005) 'Performance-Based Building and Innovation: Balancing Client and Industry Needs', *Building Research and Information*, 33 (2): 142–148.

Sexton, M., Barrett, P. and Aouad, G. (1999) *Diffusion Mechanisms for Construction Innovation and Research into Small to Medium Sized Construction Firms*, CRISP, London.

Sexton, M., Goulding, J., Zhang, X., Kagioglou, M, Aouad, G., Cooper, R. and Barrett, P. (2005) 'The Role of the HyCon Design-support Tool in Elevating Hybrid Concrete as a Design Option for Structural Frames', *Engineering, Construction and Architectural Management*, 12 (6): 568–587.

Sexton, M., Barrett, P. and Aouad, G. (2006) 'Motivating Small Construction Companies to Adopt New Technology', *Building Research and Information*, 34 (1): 11–22.

Sherer, P.D. (1995) 'Leveraging Human Assets in Law Firms: Human Capital Structures and Organizational Capabilities', *Industrial and Labor Relations Review*, 48 (4): 671–691.

Simon, E. (1944) *The Placing and Management of Building Contracts*, HMSO, London.

Slater, S. and Narver, J.C. (1994) 'Does Competitive Environment Moderate the Market Orientation–Performance Relationship?', *Journal of Marketing*, 58: 46–55.

Slaughter, S.E. (1998) 'Models of Construction Innovation', *Journal of Construction Engineering and Management*, 124 (3): 226–231.

SME Statistics (2002) *UK SME Statistics 2001*. Available at http://www.sbs.gov.uk (accessed May 2004).

Smith, A. and Whittaker, J. (1998) 'Management Development in SMEs: What Needs to be Done?', *Journal of Small Business and Enterprise Development*, 5 (2): 176–185.

Snow, C. and Hrebiniak, L.G. (1980) 'Strategy, Distinct Competence, and Organizational Performance', *Administrative Science Quarterly*, 25: 317–336.

Spender, J.C. (1989) *Industry Recipes: An Inquiry into the Nature and Sources of Management Judgement*, Basil Blackwell, Cambridge, MA.

Stanworth, J. and Gray, C. (eds) (1991) *Bolton 20 Years On*, Paul Chapman, London.

Steinmuller, W.E. (2000) 'Does the United States Need a Technology Policy?', in C. Howes and A. Singh (eds) *Competitiveness Matters: Industry and Economic Performance in the U.S.*, Ann Arbor, MI, University of Michigan Press.

Storey, D.J. (1986) 'The Economics of Small Businesses: Some Implications for Regional Economic Development', in A. Amin and R. Goddard (eds) *Technological Change, Industrial Restructuring and Regional Development*, Allen & Unwin, London.

Storey, D.J. (1998) *Understanding the Small Business Sector*, International Thomson Business Press, London.

Storey, D.J. and Cressy, R. (1995) *Small Business Risk: A Firm and Bank Perspective*, Working Paper, SME Centre, Warwick Business School, Warwick, UK.

Storey, D.J. and Sykes, N.G. (1996) 'Uncertainty, Innovation and Management', in P. Burns and J. Dewhurst (eds) *Small Business and Entrepreneurship*, Macmillan, London.

Sung, T.K. and Gibson, D.V. (2000) 'Knowledge and Technology Transfer: Levels and Key Factors', *Proceedings of the Fourth International Conference on Technology Policy and Innovation*, Brazil, August.

Sveiby, K.E. (1997) *The New Organizational Wealth: Managing and Measuring Knowledge-based Assets*, Berrett-Koehler, San Francisco, CA.

Sveiby, K.E. and Lloyd, T. (1987) *Managing Knowledge*, Bloomsbury, London.

Tatum, C.B. (1984) 'What Prompts Construction Innovation?', *Journal of Construction Engineering and Management*, 100 (3): 311–323.

Tatum, C.B. (1986) 'Potential Mechanisms for Construction Innovation', *Journal of Construction Engineering and Management*, 112 (2): 178–191.

Tatum, C.B. (1989) 'Organising to Increase Innovation in Construction Firms', *Journal of Construction Engineering and Management*, 115 (4): 602–617.

Teece, D.J., Pisano, G. and Shuen, A. (1997) 'Dynamic Capabilities and Strategic Management', *Strategic Management Journal*, 18: 509–533.

Thomas, G. and Bone, R. (2000) *Innovation at the Cutting Edge: The Experience of Three Infrastructure Projects – CIRIA Funders Report FR/CP/79*, CIRIA (Construction Industry Research and Information Association), London.

Thompson, J.D. (1967) *Organizations in Action: Social Science Bases of Administrative Theory*, McGraw-Hill, New York.

Thompson, V.A. (1965) 'Bureaucracy and Innovation', *Administrative Science Quarterly*, 5: 1–120.

Tidd, J., Bessant, J. and Pavitt, K. (1997) *Managing Innovation: Integrating Technological, Market and Organizational Change*, Wiley, Chichester, UK.

Toole, M.T. (1998) 'Uncertainty and Home Builders' Adoption of Technological Innovations', *Journal of Construction Engineering and Management*, 124 (4): 323–332.

Treacy, M. and Wiersema, F. (1995) *The Discipline of Market Leaders*, HarperCollins, London.

Urabe, K. (1988) *Innovation and Management*, Walter de Gruyter, New York.

Utterback, J.M. (1994) *Mastering the Dynamics of Innovation*, Harvard Business School Press, Boston, MA.

Van de Ven, A.H. (1986) 'Central Problems in the Management of Innovation', *Management Science*, 32: 590–607.

Van de Ven, A.H., Polley, D.E., Garud, R. and Venkataraman, S. (1999) *The Innovation Journey*, Oxford University Press, Oxford.

Verona, G. (1999) 'A Resource-based View of Product Development', *Academy of Management Review*, 24 (1): 132–142.

Veshosky, D. (1998) 'Managing Innovation Information in Engineering and Construction Firms', *Journal of Management in Engineering*, 14 (1): 58–66.

Walker, A. (1989) *Project Management in Construction*, BSP Professional Books, Oxford.

Walsh, J.P. and Ungson, G.R. (1991) 'Organizational Memory', *Academy of Management Review*, 16: 57–91.

Welsh, J.A. and White, J.F. (1981) 'A Small Business is Not a Little Big Business', *Harvard Business Review*, 59 (4): 18–32.

Wheatley, M.J. (1992) *Leadership and the New Science*, Berrett-Koehler, San Francisco, CA.

Whittaker, J., Smith, A., Boocock, G. and Loan-Clarke, J. (1997) *Management NVQs and Small and Medium Sized Enterprises*, Project Report for MCI and DfEE, March, London.

Wilson, I. (1986) 'The Strategic Management of Technology: Corporate Fad or Strategic Necessity?', *Long Range Planning*, 19 (2): 21–22.

Winch, G.M. (1998) 'Zephyrs of Creative Destruction: Understanding the Management of Innovation in Construction', *Building Research and Information*, 26 (5): 268–279.

Winter, S.G. (2000) 'The Satisfying Principle in Capability Learning', *Strategic Management Journal*, 21 (10–11): 981–996.

Woodcock, D.J., Mosey, S.P. and Wood, T.B.W. (2000) 'New Product Development in British SMEs', *European Journal of Innovation Management*, 3 (4): 212–221.

Yasai-Ardekani, M. (1986) 'Structural Adaptations to Environments', *Academy of Management Review*, 11: 9–21.

Zahra, S. (1991) 'Predictors and Financial Outcomes of Corporate Entrepreneurship: An Exploratory Study', *Journal of Business Venturing*, 6: 259–285.

Zaltman, J., Duncan, R. and Holbek, J. (1973) *Innovations and Organizations*, Wiley, New York.

Index

Printed and bound by CPI Group (UK) Ltd, Croydon, CR0 4YY

22/10/2024

01777613-0018